Introduction to the
MATHEMATICAL THEORY
OF CONTROL PROCESSES

VOLUME I
Linear Equations and Quadratic Criteria

This is Volume 40 in
MATHEMATICS IN SCIENCE AND ENGINEERING
A series of monographs and textbooks
Edited by RICHARD BELLMAN, *University of Southern California*

The complete listing of the books in this series is available from the Publisher upon request.

Introduction to the
MATHEMATICAL THEORY
OF CONTROL PROCESSES

Richard Bellman

DEPARTMENTS OF MATHEMATICS,
ELECTRICAL ENGINEERING, AND MEDICINE
UNIVERSITY OF SOUTHERN CALIFORNIA
LOS ANGELES, CALIFORNIA

VOLUME I
Linear Equations and Quadratic Criteria

1967

A C A D E M I C   P R E S S   New York and London

ACADEMIC PRESS INC.
111 Fifth Avenue, New York, New York 10003

United Kingdom Edition published by
ACADEMIC PRESS INC. (LONDON) LTD
Berkeley Square House, London W.1

LIBRARY OF CONGRESS CATALOG CARD NUMBER: 67–23153

Second Printing, 1972

PRINTED IN THE UNITED STATES OF AMERICA

1342096

INTRODUCTION

A new mathematical discipline has emerged from the bustling scientific activity of the last fifteen years, the theory of control processes. Its mathematical horizons are unlimited and its applications increase in range and importance with each passing year. Consequently, it is reasonable to believe that introductory courses in control theory are essential for training the modern graduate student in pure and applied mathematics, engineering, mathematical physics, economics, biology, operations research, and related fields.

The extremely rapid growth of the theory, associated intimately with the continuing trend toward automation, makes it imperative that courses of this nature rest upon a broad basis. In this first volume, we wish to cover the fundamentals of the calculus of variations, dynamic programming, discrete control processes, use of the digital computer, and functional analysis.

The basic problem in attempting to present this material is to provide a rigorous, yet elementary, account geared to the abilities of a student who can be safely presumed to possess only an undergraduate training in calculus and differential equations, together with a course in matrix theory. We can achieve our objectives provided that the order of presentation in the text is carefully scheduled, and that we focus upon the appropriate problems.

In Chapters 1 and 3, we discuss in a purely expository fashion a number of topics in the study of systems and stability which furnish the cultural background for the purely analytic theory. Chapter 2 provides a review of the properties of second-order linear differential equations that we employ in Chapters 4 and 5.

In Chapter 4, we begin the mathematical exposition with the problem of minimizing the quadratic functional

$$J(u, v) = \int_0^T (u^2 + v^2) \, dt,$$

where u and v are related by the linear differential equation

$$\frac{du}{dt} = au + v, \qquad u(0) = c.$$

To treat this problem rigorously in a simple fashion, we pursue a roundabout course. First we employ a formal method to obtain the Euler equation, a fundamental necessary condition. Then we show by means of a simple direct calculation that this equation has a unique solution, and that this solution provides the desired minimum value.

The analytic techniques used are carefully selected so as to generalize to the multidimensional problems treated in Chapter 7.

Chapter 5 contains an introductory account of dynamic programming with applications to the problems resolved in the previous chapter. Using the explicit solutions obtained from the calculus of variations, we can easily demonstrate that the methods of dynamic programming furnish valid results. Once again we follow the same path: first a formal derivation of the desired results and then a proof that the results obtained in this fashion are correct.

In this chapter, we also give a brief account of discrete control processes. The technique of dynamic programming is particularly well suited to this important type of process and there is no difficulty in applying dynamic programming in a rigorous fashion from the very beginning.

Having described the dual approaches of the calculus of variations and dynamic programming in the scalar case, we are ready to tackle the multidimensional optimization problems. In order to do this, it is necessary to employ some vector-matrix notation, together with some elementary properties of matrices. In Chapter 6, we review the application of matrix theory to linear differential equations of the form

$$\frac{dx_i}{dt} = \sum_{j=1}^N a_{ij} x_j, \qquad x_i(0) = c_i, \qquad i = 1, 2, \ldots, N.$$

We are now ready to study the problem of minimizing the quadratic functional

$$J(x, y) = \int_0^T [(x, x) + (y, y)] \, dt,$$

where x and y are related by means of the differential equation

$$\frac{dx}{dt} = Ax + y, \qquad x(0) = c.$$

Here x and y are N-dimensional vectors.

A straightforward generalization of the methods of Chapters 4 and 5, aided and abetted by the power of matrix analysis, enables us in Chapters 7 and 8 to give, respectively, a treatment by means of the calculus of variations and dynamic programming.

What is new here, and completely masked in the scalar case, is the difficult problem of obtaining an effective numerical solution of high-dimensional optimization problems. In each chapter there are short discussions of some of the questions of numerical analysis involved in this operation.

The concluding chapter, Chapter 9, contains an account of a direct attack upon the original variational problems by means of functional analysis. This is the only point where the book is not self-contained, since we require the rudiments of Hilbert space theory for the first part of the chapter, and a bit more for the final part of the chapter.

At the ends of the various sections inside the chapters, there are a number of exercises that illustrate the text. At the ends of the chapters, we have placed a number of problems that are either of a more difficult nature or that relate to topics that we have reluctantly omitted from the text in order to avoid any undue increase in the mathematical level. References to original sources and to more detailed results are given with these exercises. Much of this material can be used for classroom presentation if the instructor feels the class is at the appropriate level.

The second volume in this series will be devoted to a discussion of the problems associated with the analytic and computational treatment of more general optimization problems of deterministic type involving nonlinear differential equations and functionals of more general type.

The third volume will begin the treatment of stochastic control processes with emphasis upon Markovian decision processes and linear equations with quadratic criteria.

Los Angeles, California   Richard Bellman

CONTENTS

Chapter 3. **Stability and Control**

Chapter 4. **Continuous Variational Processes: Calculus of Variations**

Chapter 5. **Dynamic Programming**

Chapter 6. Review of Matrix Theory and Linear Differential Equations

Chapter 7. Multidimensional Control Processes via the Calculus of Variations

Chapter 8.   Multidimensional Control Processes via Dynamic Programming

Chapter 9.   Functional Analysis

Contents

Introduction to the
MATHEMATICAL THEORY
OF CONTROL PROCESSES

VOLUME I
Linear Equations and Quadratic Criteria

I

WHAT IS CONTROL THEORY?

1.1. Introduction

This is the first of a series of volumes devoted to an exposition of a new mathematical theory, the theory of control processes. Despite the fact that no attempt will be made to discuss any of the many possible applications of the methods and concepts that are presented in the following pages, it is worthwhile to provide the reader with some idea of the scientific background of the mathematical theory. There are many reasons for doing this. In the first place, it is very easy for the student to become overwhelmed with the analytic formalism unless he has some clear idea of the general direction that is being pursued; in the second place, control theory is still a rapidly developing field with the greater part yet to come. We prefer to have the reader strike out boldly into the new domains rather than trudge in well-ploughed fields, and we feel strongly that the most significant and exciting mathematics will arise in the study of the most important of contemporary areas requiring control and decisionmaking.

Finally, there is the matter of intellectual curiosity. A person who claims the distinction of being well-educated should know the origins and applications of his field of specialization.

1.2. Systems

Control theory centers about the study of systems. Indeed, one might describe control theory as the care and feeding of systems. Intuitively, we can consider a system to be a set of interacting components subject to various inputs and producing various outputs. There is little to be gained at this stage of development from any formal definition.

We are all familiar with many different types of systems: mechanical systems such as clocks, electrical systems such as radios, electro-mechanical-chemical systems such as automobiles, industrial systems such as factories, medical systems such as hospitals, educational systems such as universities, animate systems such as human beings, information processing systems such as digital computers, and many others that could be catalogued. This brief list, however, is sufficient to emphasize the fact that one of the most profound concepts in current culture is that of a " system." The ramifications of this product of civilization will occupy the intellectual world for a long time to come. In particular, it is an unlimited source of fertile mathematical ideas.

1.3. Schematics

It is often a convenient guide to both intuition and exposition to conceive of a system in terms of a block diagram. A first attempt is:

Input ⟶ | System | ⟶ Output

Frequently, this pictogram is much too crude, and we prefer the following diagram to indicate the fact that a system usually has a variety of inputs and outputs.

I_1 ⟶ | | ⟶ O_1

I_2 ⟶ | System | ⟶ O_2

⋮

I_M ⟶ | | ⟶ O_N

Diagrams of this type, in turn, must yield to still more sophisticated representations. Thus, for example, in chemotherapy, we often use the following diagram in connection with the study of the time history of a drug and its by-products injected into the bloodstream. Here, S_1 represents the heart pumping blood through the extracellular and intracellular tissues of two different organs, $[S_2, S_3]$ and $[S_4, S_5]$.

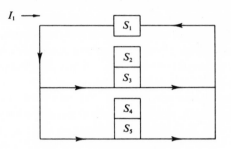

Figure 1.1

In multistage production processes, or chemical separation processes, we may employ a schematic such as:

$I_1 \to$ | S_1 | $\to O_1 \to$ | S_2 | $\to \cdots \to$ | S_N | $\to$
$I_2 \to$ | | $\to O_2 \to$ | | $\to \cdots \to$ | | $\to$

This indicates that the inputs, I_1, I_2, to subsystem S_1 produce outputs O_1, O_2 that are themselves inputs to subsystem S_2, and so on until the final output.

In the study of atmospheric physics, geological exploration, and cardiology, we frequently think in terms of diagrams such as

$I_1 \to$ | S_1 | $\to$ | | $\to$ | S_4 | $\to O_1$
                            | S_3 |
$I_2 \to$ | S_2 | $\to$ | | $\to$ | S_5 | $\to O_2$

bringing clearly to the fore that there are internal subsystems which we can neither directly examine nor influence. The associated questions of identification and control of S_3 are quite challenging.

Schematics of the foregoing nature are designed to aid our thinking. This means that we must be careful not to take them too seriously. What will be a useful representation in one case can be quite misleading in another. An electrical system that can very accurately be conceived of in terms of resistances, inductances, and capacitances, that is, lumped parameters, in one frequency range requires an entirely different representation in another frequency range.

Simplicity is essential for scientific progress, but systems are basically not simple, and the control of systems is equally not simple. Consequently constant examination of our concepts and methods is essential to make sure that we avoid both the self-fulfilling prophecy and the self-defeating simplifying assumption.

Once the mathematical die has been cast, equations assume a life of their own and can easily end by becoming master rather than servant.

1.4. Mathematical Systems

In the same vein, it cannot be too strongly emphasized that real systems possess many different conceptual and mathematical realizations. Each mathematical description of a system possesses certain advantages and disadvantages. Some are designed for the purposes of analysis of the structure of the behavior of the system over time; some are designed principally to provide numerical answers to numerical questions; some in conjunction with analog computers; some in conjunction with digital computers; some for extreme precision; some for crude approximation. Ideally, there should be mathematical formulations for all purposes. In practice, little thought has been devoted to these important types of questions.

In any case, it should constantly be kept in mind that the mathematical system is never more than a projection of the real system on a conceptual axis.

1.5. The Behavior of Systems

The behavior of systems is not always exemplary, so much the worse for the world and so much the better for the mathematician and systems engineer. In all parts of contemporary society, we observe systems that are not operating in a completely desirable fashion. We note economic systems subject to inflation and depression, human systems subject to cancer and heart disease, industrial systems subject to unproductivity and fits of unreliability, university systems subject to obsolescence, and ecological systems subject to pests and drought. These examples can be multiplied indefinitely.

What we can conclude from this sorry recital is that systems do indeed require care. They do not operate well in a complete laissez-faire climate. On the other hand, as we shall see, the cost of supervision must be kept clearly in mind. This cost may be spelled out in terms of resources, money, manpower, time, or complexity. An increase in complexity usually results in a decrease of reliability.

1.6. Improvement of the Behavior of Systems

There are several immediate ways of improving the performance of a system. To begin with, we can either build a new system, or change the inputs. In some cases, as, say, an old car, a new system may be the ideal solution. In other cases, say the human body or an economic system, replacement is not a feasible procedure.

Consequently, we shall think in terms of the more feasible program of altering the design of the system or in terms of modifying the inputs, or both. One way to do this is to observe the output of the system and to make these alterations and modifications dependent upon the deviation of the actual output from the desired output. This is the basic idea of feedback control as shown schematically in Figure 1.2.

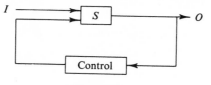

Figure 1.2

The idea is both ingenious and sound. Naturally, there is a great deal more to making it operational than the simple schematic appearing above might suggest.

1.7. More Detailed Breakdown

It is important for the mathematical models we will subsequently construct to examine the situation in finer detail. Thus, for example, the operation of observing the output is quite distinct from the operation of exerting control. We shall analyze this vital operation of obtaining information concerning the system in great detail in Volumes II and III. In this volume, we will adhere to the classical convention that observation of the output is exact, instantaneous, and free. Obviously, this is never the situation in the study of real systems. But the practicing scientist is reconciled to the fact that reality can only be approached through unreality.

At very least, we should use a more detailed schematic (see Figure 1.3).

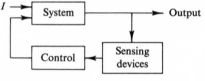

Figure 1.3

If we now take into account the fact that we will, in general, observe the system itself, as well as the output, and that decision making concerning the choice of a control action is really an operation distinct from that of exerting the control action decided upon, then the preceding figure expands. We can, for example, think in terms of the diagram shown in Figure 1.4.

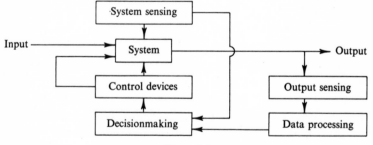

Figure 1.4

It is important to emphasize the interaction between the system and the sensing and controlling devices, as the arrows pointing in both directions are intended to indicate.

Further careful analysis of the actual operations involved in observing and controlling a system will introduce still additional decompositions into subsystems that we will not describe here. Our aim in this brief discussion is solely to indicate to the reader how much is involved in the mathematical formulation of a control process and, thus, how careful one must be not to use ready-made analytic formulations merely because they are there. This may be a reasonable excuse for mountain-climbing, but not for science.

The fact that new approaches are needed to new types of problems is a very cheering one to the research mathematician.

1.8. Uncertainty

We have already indicated the classical convention of exact, instantaneous, and free observation. As soon as we admit the existence of a lack of completely accurate measurement, or, as we shall say, uncertainty, in one part of the system, we must allow for it in every part. There are many subtle and difficult problems of intense mathematical interest that arise in connection with uncertainty, the discussion of which we will defer until Volume III.

Our object in mentioning the topic here is to remind the reader that elegant and satisfying as our present-day theory of deterministic control processes is, it remains only a rough approximation to a future, more satisfying theory of the control of real systems.

1.9. Conclusion

We have very quickly sketched in the background of the study of systems that motivates so much of the current mathematical theory of control processes. With the broad picture as a backdrop, we shall now proceed to fill in carefully the foundations of a theory of deterministic control processes. Henceforth, we shall operate completely within the secure domain of mathematical systems.

BIBLIOGRAPHY AND COMMENTS

1.1. The reader may find the following books and papers helpful in amplifying the preceding discussion.

M. Athans, "The Status of Optimal Control Theory and Applications for Deterministic Systems," *IEEE Trans. Auto. Control*, **AC-11**, 1966, pp. 580–596.

R. Bellman, *Adaptive Control Processes: A Guided Tour*, Princeton Univ. Press, Princeton, New Jersey, 1961.

R. Bellman and R. Kalaba, *Mathematical Trends in Control Theory*, Dover, New York, 1964.

P. Eykhoff, "Adaptive and Optimalizing Control Systems," *IRE Trans. Auto. Control*, **AC-5**, 1960, pp. 148–151.

I. M. Horowitz, "Plant Adaptive Systems Versus Ordinary Feedback Systems," *IRE Trans. Auto. Control*, **AC-7**, 1962, pp. 48–56.

B. Paiewonsky, "Optimal Control: A Review of Theory and Practice," *AIAA*, **3**, 1963, pp. 1985–2006.

W. C. Schultz, "Control System Performance Measures: Past, Present, Future," *IRE Trans. Auto. Control*, **AC-6**, 1961, pp. 22–35.

P. L. Simmons and H. A. Pappo, "Soviet Literature on Control Systems," *IRE Trans. Auto. Control*, **AC-5**, 1960, pp. 142–147.

L. Zadeh, "Optimality and Non-Scalar-Valued Performance Criteria," *IEEE Trans. Auto. Control*, **AC-8**, 1963, pp. 59–60.

1.3. For some of the ways in which mathematical flesh is put around the schematic skeletons of this section, see

R. Bellman, J. Jacquez, and R. Kalaba, "Some Mathematical Aspects of Chemotherapy—I: One-Organ Models," *Bull. Math. Biophys.*, **22**, 1960, pp. 181–198.

R. Bellman, J. Jacquez, and R. Kalaba, "Some Mathematical Aspects of Chemotherapy—II: The Distribution of a Drug in the Body," *Bull. Math. Biophys.*, **22**, 1960, pp. 309–322.

R. Bellman, J. Jacquez, and R. Kalaba, "Mathematical Models of Chemotherapy," *Proc. Fourth Berkeley Symposium on Mathematical Statistics and Probability*, Univ. of California Press, Berkeley, California, I, 1961, pp. 37–48.

R. Aris, *The Optimal Design of Chemical Reactors: A Study in Dynamic Programming*, Academic Press, New York, 1961.

S. M. Roberts, *Dynamic Programming in Chemical Engineering and Process Control*, Academic Press, New York, 1964.

F. S. Grodins, *Control Theory and Biological Systems*, Columbia Univ. Press, New York, 1963.

W. S. Yamamoto and J. R. Brobeck, *Physiological Controls and Regulations*, W. B. Saunders, Philadelphia, 1965.

1.7. See

R. Bellman, M. B. Friend, and L. Kurland, *A Simulation of the Initial Psychiatric Interview*, The RAND Corporation, **R-449-RC**, 1966, for a detailed discussion of this interaction in the initial psychiatric interview.

1.8. A discussion of some of the different types of uncertainty that can occur in the analysis of systems will be found in

R. Bellman, *Adaptive Control Processes: A Guided Tour*, Princeton Univ. Press, Princeton, New Jersey, 1961.

2

SECOND-ORDER LINEAR DIFFERENTIAL

AND DIFFERENCE EQUATIONS

2.1. Introduction

In this chapter, we wish to review briefly a number of results in the theory of second-order linear differential and difference equations that we will use frequently in the following chapters. The presentation will be self-contained, but brisk.

2.2. Second-Order Linear Differential Equations with Constant Coefficients

The differential equation for the function $u(t)$

$$u'' + a_1 u' + a_2 u = 0, \qquad (2.2.1)$$

where the prime denotes the derivative with respect to t, and where a_1 and a_2 are constants, has as its general solution

$$u = c_1 e^{\lambda_1 t} + c_2 e^{\lambda_2 t}, \qquad (2.2.2)$$

where λ_1 and λ_2 are the roots of the quadratic equation

$$\lambda^2 + a_1 \lambda + a_2 = 0, \qquad (2.2.3)$$

provided that λ_1 and λ_2 are distinct. Here c_1 and c_2 are arbitrary constants. If λ_1 is a multiple root, the general solution has the form

$$u = c_1 e^{\lambda_1 t} + c_2 t e^{\lambda_1 t}, \tag{2.2.4}$$

where c_1 and c_2 are again arbitrary constants. It is this property of superposition that makes linear equations an indispensable tool of analysis.

The solution is, in any case, uniquely determined by the initial conditions

$$u(0) = b_1, \qquad u'(0) = b_2. \tag{2.2.5}$$

If $\lambda_1 \neq \lambda_2$, we obtain from (2.2.2) and (2.2.5) the system of simultaneous linear algebraic equations

$$\begin{aligned} b_1 &= c_1 + c_2, \\ b_2 &= \lambda_1 c_1 + \lambda_2 c_2. \end{aligned} \tag{2.2.6}$$

Under our assumption that $\lambda_1 \neq \lambda_2$, the determinant is nonzero and there is thus a unique solution for c_1 and c_2.

If the roots coincide, we obtain the equations

$$\begin{aligned} b_1 &= c_1, \\ b_2 &= c_1 \lambda_1 + c_2, \end{aligned} \tag{2.2.7}$$

again with a unique solution.

EXERCISES

1. Show that the solution of (2.2.1) and (2.2.5) is given by

$$u = b_1 \left(\frac{e^{\lambda_1 t} + e^{\lambda_2 t}}{2} \right) + \left(b_2 - \frac{b_1}{2} (\lambda_1 + \lambda_2) \right) \left(\frac{e^{\lambda_1 t} - e^{\lambda_2 t}}{\lambda_1 - \lambda_2} \right)$$

   if $\lambda_1 \neq \lambda_2$.

2. Show that the limiting form of this solution as $\lambda_2 \to \lambda_1$ is precisely the solution obtained from (2.2.7).

3. Can one establish the validity of this limiting relation without explicit calculation? (In other words, are the solutions of (2.2.1) continuous functions of the coefficients a_1 and a_2 in any finite t-interval?)

4. What are necessary and sufficient conditions on a_1 and a_2 in order that $u(t)$ be periodic in t?

2.3. The Inhomogeneous Equation

The solution of the inhomogeneous equation

$$u'' + a_1 u' + a_2 u = f(t), \qquad u(0) = c_1, \qquad u'(0) = c_2, \qquad (2.3.1)$$

can be obtained as the sum of two solutions, $u = v + w$, where

$$\begin{aligned} v'' + a_1 v' + a_2 v &= 0, & v(0) &= c_1, & v'(0) &= c_2, \\ w'' + a_1 w' + a_2 w &= f(t), & w(0) &= 0, & w'(0) &= 0. \end{aligned} \qquad (2.3.2)$$

The solution of the second equation can readily be obtained by means of the method described in Section 2.8. However, it is worth noting that as long as the coefficients are constant, we can make use of the Laplace transform. We have

$$L(w'' + a_1 w' + a_2 w) = L(f), \qquad (2.3.3)$$

where

$$L(w) = \int_0^\infty e^{-st} w(t) \, dt, \qquad (2.3.4)$$

and the real part of s is taken sufficiently large. Integrating by parts and using the initial conditions of (2.3.2), we have

$$L(w) = \frac{L(f)}{s^2 + a_1 s + a_2}. \qquad (2.3.5)$$

Hence,

$$w(t) = \int_0^t k(t - t_1) f(t_1) \, dt_1, \qquad (2.3.6)$$

where $k(t)$ is determined by the relation

$$L(k) = \frac{1}{s^2 + a_1 s + a_2}. \qquad (2.3.7)$$

If the characteristic roots λ_1 and λ_2 are distinct, we see that

$$\frac{1}{s^2 + a_1 s + a_2} = \frac{1}{(\lambda_1 - \lambda_2)} \left[\frac{1}{(s - \lambda_1)} - \frac{1}{(s - \lambda_2)} \right], \qquad (2.3.8)$$

whence

$$k(t) = \frac{e^{\lambda_1 t} - e^{\lambda_2 t}}{\lambda_1 - \lambda_2}.$$ (2.3.9)

If λ_1 is a multiple root, we see that

$$k(t) = te^{\lambda_1 t}.$$ (2.3.10)

EXERCISES

1. Write the solution of $u'' - u = f(t)$, $u(0) = 1$, $u'(0) = 0$.
2. Write the solution of $u'' + u = f(t)$, $u(0) = 1$, $u'(0) = 0$.
3. Show that a necessary and sufficient condition that all solutions of $u'' + a_1 u' + a_2 u = 0$ approach zero as $t \to \infty$ is that the roots of $r^2 + a_1 r + a_2 = 0$ have negative real parts.
4. What is a necessary and sufficient condition for this?

2.4. Two-Point Boundary Conditions

In many problems of interest, in particular in many areas of mathematical physics and in the calculus of variations, as we shall see in Chapter 3, we encounter equations subject to two-point boundary conditions rather than the simpler initial-value problems treated in the foregoing sections. Consider, for example,

$$u'' + a_1 u' + a_2 u = 0, \qquad u(0) = b_1, \qquad u(T) = b_2.$$ (2.4.1)

Starting with the general solution of the differential equation

$$u = c_1 e^{\lambda_1 t} + c_2 e^{\lambda_2 t},$$ (2.4.2)

assuming for the moment that $\lambda_1 \neq \lambda_2$, we obtain from the two boundary conditions the set of simultaneous algebraic equations

$$\begin{aligned} c_1 + c_2 &= b_1, \\ c_1 e^{\lambda_1 T} + c_2 e^{\lambda_2 T} &= b_2. \end{aligned}$$ (2.4.3)

There is a unique solution provided that the determinant

$$\Delta(T) = \begin{vmatrix} 1 & 1 \\ e^{\lambda_1 T} & e^{\lambda_2 T} \end{vmatrix} = e^{\lambda_2 T} - e^{\lambda_1 T} \neq 0.$$ (2.4.4)

If λ_1 and λ_2 are real and distinct, the quantity $e^{\lambda_1 T} - e^{\lambda_2 T}$ is nonzero for $T > 0$. If, however, the characteristic roots are complex conjugates,

$$\lambda_1, \lambda_2 = -a_3 \pm ia_4, \qquad (2.4.5)$$

$a_4 > 0$, then we see that

$$\Delta(T) = e^{\lambda_1 T} - e^{\lambda_2 T} = e^{-a_3 T}(e^{ia_4 T} - e^{-ia_4 T}), \qquad (2.4.6)$$

and hence that the determinant is zero for $T = \pi/a_4 + 2n\pi/a_4$, $n = 1, 2, \ldots$.

The fact to emphasize is that two-point boundary conditions are quite different from initial conditions as far as determining solutions is concerned. Even if $\Delta(T) = 0$, the values b_1 and b_2 may be such that a solution to (2.4.1) exists. In this case, there is an infinite set of solutions. Consider, for example,

$$u'' + u = 0, \qquad u(0) = 0, \qquad u(\pi) = 0. \qquad (2.4.7)$$

All functions of the form $u = c_1 \sin t$ are solutions where c_1 is an arbitrary constant.

EXERCISES

1. Under what conditions do the conditions $u(0) = c_1$, $\int_0^T u \, dt_1 = c_2$, uniquely determine a solution of $u'' + a_1 u' + a_2 u = 0$?
2. Under what conditions do the conditions

$$\int_0^T e^{a_3 t_1} u \, dt_1 = c_1, \qquad \int_0^T e^{a_4 t_1} u \, dt_1 = c_2$$

uniquely determine a solution of $u'' + a_1 u' + a_2 u = 0$?

2.5. First-Order Linear Differential Equations with Variable Coefficients

Let us next consider the general linear first-order differential equation

$$\frac{du}{dt} = p(t)u + q(t), \qquad u(0) = c. \qquad (2.5.1)$$

Using the integrating factor $\exp(-\int_0^t p(t_1)\,dt_1)$, we see that this can be written

$$\frac{d}{dt}\left(u\,\exp\left(-\int_0^t p\,dt_1\right)\right) = \exp\left(-\int_0^t p\,dt_1\right)q(t). \qquad (2.5.2)$$

Hence, the general solution of (2.5.1) is

$$u = c\,\exp\left(\int_0^t p\,dt_1\right) + \int_0^t \exp\left(\int_{t_1}^t p\,dt_1\right)q(t_1)\,dt_1. \qquad (2.5.3)$$

EXERCISES

1. How many solutions of $u' + au = 1$ can satisfy the condition that $\lim_{t\to\infty} u(t)$ exists? Consider separately the three cases $a > 0$, $a < 0$, $a = 0$.

2. Consider the same question for $u'' - au = 1$.

3. Consider the same question for $u' + a(t)u = f(t)$ under the assumption that the limits of $a(t)$ and $f(t)$ exist as $t \to \infty$. Take the case where $a(t)$ is constant first.

4. Let u be a function satisfying the differential inequality $u' \le p(t)u + q(t)$, $u(0) = c$, and let v satisfy the equation $v' = p(t)v + q(t)$, $v(0) = c$. Show that $u \le v$ for $t \ge 0$. (*Hint:* The differential inequality is equivalent to the equation $u' = p(t)u + q(t) - r(t)$, $u(0) = c$, where $r(t) \ge 0$ for $t \ge 0$.)

5. Show that $u \le c + \int_0^t uv\,ds$ for $t \ge 0$ with $u, v, c \ge 0$, c a constant, implies that

$$u \le c\,\exp\left(\int_0^t v\,ds\right) \qquad \text{for} \quad t \ge 0.$$

(*Hint:* Convert this into a differential inequality for $w = \int_0^t uv\,ds$.) This is the "fundamental lemma" of the theory of differential equations which plays a basic role in existence and uniqueness theory and stability theory.

2.6. The Riccati Equation

A first-order nonlinear differential equation

$$\frac{du}{dt} = p_1(t) + p_2(t)u + p_3(t)u^2, \qquad u(0) = c, \qquad (2.6.1)$$

the Riccati equation, will play an important role in what follows in connection with our use of dynamic programming in Chapter 4.

If p_1, p_2, p_3 are constant, we see that (2.6.1) can be solved by separation of variables. A more general way of approaching this equation is to note that the linear equation

$$v'' + a_1 v' + a_2 v = 0, \qquad (2.6.2)$$

where a_1 and a_2 are functions of t, goes over into an equation of the form of (2.6.1) upon the change of variable

$$v = \exp\left(\int_0^t u \, dt_1\right). \qquad (2.6.3)$$

We have

$$v' = uv,$$
$$v'' = uv' + u'v = (u' + u^2)v. \qquad (2.6.4)$$

Hence, if $v \neq 0$, we obtain

$$u' + u^2 + a_1 u + a_2 = 0. \qquad (2.6.5)$$

Conversely, we can pass from (2.6.5) to a linear equation upon setting $u = v'/v$.

EXERCISES

1. Solve $u' = 1 + u^2$, $u(0) = c$. Does the solution exist for all $t > 0$? What is the critical value of t?
2. Solve $u' = 1 - u^2$, $u(0) = c$. Does the solution exist for all $t > 0$? Discuss the behavior of the solution as $t \to \infty$ for $|c| < 1$, $c = \pm 1$, $c > 1$, $c < -1$.
3. Show that $u(t)^{-1}$ satisfies a Riccati differential equation when u does and that generally $w = (b_1(t)u + b_2(t))/(b_3(t)u + b_4(t))$ satisfies a Riccati differential equation when u does.
4. Show that $u^2 = \max_w (2uw - w^2)$.
5. Consider the equation $u' = q(t) + u^2$, $u(0) = c$. From the foregoing, it may be written

$$u' = \max_w (2uw - w^2 + q(t)). \qquad (1)$$

Consider the linear differential equation

$$U' = 2Uw - w^2 + q(t), \qquad U(0) = c, \tag{2}$$

where $w(t)$ is now an arbitrary function of t. Show that $u \geq U$ for $t \geq 0$. (*Hint:* (1) is equivalent to $u' \geq 2uw - w^2 + q(t)$ for all $w(t)$, or $u' = 2uw - w^2 + q(t) + p(t)$, where $p(t) \geq 0$.)

6. Conclude that $u = \max_w U$, and thus that the solution of $u' = q(t) + u^2$, $u(0) = c$, may be written in the form

$$u = \max_w \left[c \exp\left(2 \int_0^t w \, dt_1\right) + \int_0^t \exp\left(2 \int_{t_1}^t w \, dt_2\right) [q(t_1) - w^2] \, dt_1 \right].$$

7. How do we obtain lower bounds for $u(t)$ from the foregoing?
8. Obtain upper-bounds by considering the equation for u^{-1}.
9. Use the identity $u^n = \max_w [nuw^{n-1} - (n-1)w^n]$, $n > 1$, to obtain a corresponding representation for the solution of $u' = u^n + q(t)$, $u(0) = c, n > 1$.
10. Use the identity $g(u) = \max_w [g(w) + (u - w)g'(w)]$ if $g''(u) > 0$ to obtain a corresponding representation for the solution of $u' = g(u) + q(t)$, $u(0) = c$.

2.7. Linear Equations with Variable Coefficients

Let us now consider the case where the coefficients a_1 and a_2 are functions of t,

$$u'' + a_1(t)u' + a_2(t)u = 0. \tag{2.7.1}$$

Let u_1 and u_2 be the principal solutions, the solutions determined by the initial conditions

$$\begin{aligned} u_1(0) &= 1, & u_1'(0) &= 0, \\ u_2(0) &= 0, & u_2'(0) &= 1. \end{aligned} \tag{2.7.2}$$

The solution of (2.7.1) subject to the initial conditions

$$u(0) = c_1, \qquad u'(0) = c_2, \tag{2.7.3}$$

is then given by

$$u = c_1 u_1 + c_2 u_2. \tag{2.7.4}$$

For our further purposes, let us introduce the Wronskian of the two functions u and v,

$$W(u, v) = \begin{vmatrix} u & v \\ u' & v' \end{vmatrix}. \tag{2.7.5}$$

To compute the Wronskian of u_1 and u_2, we observe that

$$\frac{d}{dt} W(u_1, u_2) = \begin{vmatrix} u_1 & u_2 \\ u_1'' & u_2'' \end{vmatrix} = \begin{vmatrix} u_1 & u_2 \\ -a_1 u_1' - a_2 u_1 & -a_1 u_2' - a_2 u_2 \end{vmatrix}$$

$$= -a_1(t)\begin{vmatrix} u_1 & u_2 \\ u_1' & u_2' \end{vmatrix} = -a_1(t)W(u_1, u_2). \tag{2.7.6}$$

Hence,

$$W(u_1, u_2) = \exp\left(-\int_0^t a_1(t_1)\, dt_1\right) W(u_1(0), u_2(0))$$

$$= \exp\left(-\int_0^t a_1(t_1)\, dt_1\right). \tag{2.7.7}$$

(Jacobi.)

EXERCISES

1. If $u'' + a_2(t)u = 0$, show that $W(u_1, u_2) = 1$ for $t \geq 0$.
2. Hence, show that $\sin^2 t + \cos^2 t = 1$ and that $\cosh^2 t - \sinh^2 t = 1$.
3. Show that the change of variable

$$u = v \exp\left(-\frac{1}{2}\int_0^t a_1(t_1)\, d t_1\right)$$

yields the equation $v'' + (a_2 - a_1'/2 - a_1^2/4)v = 0$ if u satisfies the equation $u'' + a_1 u' + a_2 u = 0$.

2.8. The Inhomogeneous Equation

Let us now consider the inhomogeneous equation

$$u'' + a_2(t)u = f(t), \qquad u(0) = c_1, \qquad u'(0) = c_2. \tag{2.8.1}$$

As before, it is sufficient to consider the case $c_1 = c_2 = 0$ and we can take $a_1(t) = 0$ without loss of generality.

To obtain the solution, we use the following device, the method of the adjoint operator. This is a fundamental method in the theory of linear functional equations, which we will use repeatedly in what follows.

Let $v(t)$ be a function that we will choose conveniently in a moment. Then

$$\int_0^t v(t_1)(u''(t_1) + a_2(t_1)u(t_1))\, dt_1 = \int_0^t v(t_1)f(t_1)\, dt_1. \qquad (2.8.2)$$

Integrating by parts repeatedly, we have

$$u'(t)v(t) - u(t)v'(t) + \int_0^t u(t_1)(v''(t_1) + a_2(t_1)v(t_1))\, dt_1 = \int_0^t v(t_1)f(t_1)\, dt_1. \qquad (2.8.3)$$

Choose $v(t_1)$ to be the solution of

$$v''(t_1) + a_2(t_1)v(t_1) = 0, \qquad 0 \le t_1 \le t, \qquad (2.8.4)$$

satisfying the initial conditions

$$v(t) = 0, \qquad v'(t) = -1. \qquad (2.8.5)$$

Write

$$v = b_1 u_1 + b_2 u_2, \qquad (2.8.6)$$

where b_1 and b_2 are to be determined using (2.8.5) and u_1, u_2 are the principal solutions. We have

$$0 = b_1 u_1(t) + b_2 u_2(t),$$
$$-1 = b_1 u_1'(t) + b_2 u_2'(t). \qquad (2.8.7)$$

Hence,

$$b_1 = \frac{\begin{vmatrix} 0 & u_2(t) \\ -1 & u_2'(t) \end{vmatrix}}{\begin{vmatrix} u_1(t) & u_2(t) \\ u_1'(t) & u_2'(t) \end{vmatrix}} = u_2(t), \qquad b_2 = \frac{\begin{vmatrix} u_1(t) & 0 \\ u_1'(t) & -1 \end{vmatrix}}{\begin{vmatrix} u_1(t) & u_2(t) \\ u_1'(t) & u_2'(t) \end{vmatrix}} = -u_1(t), \qquad (2.8.8)$$

using the fact that $W(u_1, u_2) = 1$ for $t \ge 0$. Thus,

$$v(t_1) = u_2(t)u_1(t_1) - u_1(t)u_2(t_1) \qquad (2.8.9)$$

for $0 \le t_1 \le t$. Hence, returning to (2.8.3), we obtain the desired representation,

$$u(t) = \int_0^t [u_2(t)u_1(t_1) - u_1(t)u_2(t_1)]f(t_1) \, dt_1. \qquad (2.8.10)$$

EXERCISES

1. Obtain the solution of $u'' \pm a^2 u = f(t)$, $u(0) = u'(0) = 0$ in this fashion described above.
2. (Lagrange Variation-of-Parameters Method.) To solve (2.8.1), write $u = w_1 u_1 + w_2 u_2$, where w_1 and w_2 are functions of t. Impose the condition that $u' = w_1 u_1' + w_2 u_2'$ and then that u satisfy (2.8.1) and thus determine w_1 and w_2.
3. Use this device to find a second solution in the form $u_2 = vu_1$ if u_1 is given.
4. Use the Wronskian relation $W(u_1, u_2) = 1$ for this purpose.

2.9. Green's Function

Let us now consider the equation

$$u'' + a_2(t)u = f(t), \qquad u(0) = 0, \qquad u(T) = 0. \qquad (2.9.1)$$

The general solution of

$$u'' + a_2(t)u = f(t) \qquad (2.9.2)$$

has the form

$$u = c_1 u_1 + c_2 u_2 + \int_0^t q(t, t_1)f(t_1) \, dt_1, \qquad (2.9.3)$$

where we write

$$q(t, t_1) = u_2(t)u_1(t_1) - u_2(t_1)u_1(t). \qquad (2.9.4)$$

The condition $u(0) = 0$ shows that $c_1 = 0$. The condition $u(T) = 0$ yields the equation

$$c_2 u_2(T) + \int_0^T q(T, t_1)f(t_1) \, dt_1 = 0. \qquad (2.9.5)$$

Hence,

$$c_2 = -\frac{1}{u_2(T)} \int_0^T q(T, t_1) f(t_1)\, dt_1,$$
(2.9.6)

provided that $u_2(T) \neq 0$.

The condition $u_2(T) \neq 0$ is equivalent to the condition that (2.9.1) has a unique solution. Using (2.9.6), we have

$$u = -\frac{u_2(t)}{u_2(T)} \int_0^T q(T, t_1) f(t_1)\, dt_1 + \int_0^t q(t, t_1) f(t_1)\, dt_1$$

$$= \int_0^t \left[q(t, t_1) - \frac{u_2(t)}{u_2(T)} q(T, t_1) \right] f(t_1)\, dt_1$$

$$+ \int_t^T \left[\frac{-u_2(t) q(T, t_1)}{u_2(T)} \right] f(t_1)\, dt_1$$

$$= \int_0^T k(t, t_1) f(t_1)\, dt_1,$$
(2.9.7)

where

$$k(t, t_1) = q(t, t_1) - \frac{u_2(t)}{u_2(T)} q(T, t_1), \qquad 0 \leq t_1 \leq t,$$

$$= -\frac{u_2(t) q(T, t_1)}{u_2(T)}, \qquad t \leq t_1 \leq T.$$
(2.9.8)

Using the expression for $q(t, t_1)$ given in (2.9.4), we see that

$$q(t, t_1) - \frac{u_2(t) q(T, t_1)}{u_2(T)} = [(u_2(t) u_1(t_1) - u_2(t_1) u_1(t)) u_2(T)$$

$$- (u_2(T) u_1(t_1) - u_2(t_1) u_1(T)) u_2(t)]/u_2(T)$$

$$= \frac{[u_2(t_1) u_1(T) u_2(t) - u_2(t_1) u_1(t) u_2(T)]}{u_2(T)}$$

$$= \frac{u_2(t_1)[u_1(T) u_2(t) - u_1(t) u_2(T)]}{u_2(T)}.$$
(2.9.9)

It follows from this that

$$k(t, t_1) = k(t_1, t),$$
(2.9.10)

a most important reciprocity relation. The function $k(t, t_1)$ is called the Green's function associated with the equation and the boundary conditions.

EXERCISES

1. Determine the Green's functions associated with:

(a)	$u'' = f,$	$u(0) = u(T) = 0,$
(b)	$u'' = f,$	$u(0) = 0, \quad u'(T) = 0,$
(c)	$u'' + a^2 u = f,$	$u(0) = u(T) = 0,$
(d)	$u'' + a^2 u = f,$	$u(0) = 0, \quad u'(T) = 0,$
(e)	$u'' - a^2 u = f,$	$u(0) = u(T) = 0,$
(f)	$u'' - a^2 u = f,$	$u(0) = 0, \quad u'(T) = 0,$

and verify the symmetry in all cases. For what T-intervals is the Green's function nonnegative?

2. Consider the equation $u'' - a^2 u = 0$. Is the condition $u_2(T) \neq 0$ automatically fulfilled?

3. Determine the Green's function associated with $u'' \pm a^2 u = f(t)$, $u(0) = 0$, $\int_0^T u \, dt = 0$.

2.10. Linear Systems

We will have occasion to deal with linear systems of differential equations of the form

$$\frac{du}{dt} = a_1 u + a_2 v, \qquad u(0) = c_1,$$

$$\frac{dv}{dt} = b_1 u + b_2 v, \qquad v(0) = c_2,$$

(2.10.1)

where the coefficients are functions of t. Consider first the case where the coefficients are constants. A direct approach to the solution is to eliminate u or v and obtain a second-order linear differential equation for the remaining variable. A way of doing this which avoids any examination of the coefficients a_1, a_2, b_1, b_2 is the following. Write (2.10.1) as

$$(D - a_1)u = a_2 v,$$
$$(D - b_2)v = b_1 u, \tag{2.10.2}$$

where $D = d/dt$. To eliminate v, we write

$$(D - b_2)(D - a_1)u = a_2(D - b_2)v,$$
$$a_2(D - b_2)v = a_2 b_1 u. \tag{2.10.3}$$

Then we obtain the second-order differential equation

$$(D - b_2)(D - a_1)u = a_2 b_1 u, \tag{2.10.4}$$

with the initial conditions

$$u(0) = c_1, \qquad u'(0) = a_1 c_1 + a_2 c_2. \tag{2.10.5}$$

The most elegant, and simultaneously most efficient, way to treat linear systems of differential equations is by means of matrix theory, as we indicate in Chapter 6. Consequently, we shall not spend too much time on these matters here.

For our present purposes, all we need is the superposition of solutions plus the uniqueness in order to conclude the general form of the solution of (2.10.1).

EXERCISES

1. Examine the possibility of solving the equation of (2.10.1) subject to the two-point boundary conditions $u(0) = c_1$, $u(T) = c_2$.
2. Obtain particular solutions of $u' = a_1 u + a_2 v$, $v' = b_1 u + b_2 v$, of the form $u = e^{\lambda t} c_3$, $v = e^{\lambda t} c_4$.
3. Combine particular solutions of this type to obtain the solution of (2.10.1).
4. Eliminate v from (2.10.1) in the case where a_1, a_2, b_1, b_2 depend on t.

2.11. Difference Equations

Analogous results hold for linear difference equations

$$u_{n+2} + a_1 u_{n+1} + a_2 u_n = 0. \tag{2.11.1}$$

We are interested here only in the case of constant coefficients, a_1 and a_2. Starting with the fact that the general solution of (2.11.1) has the form

$$u_n = c_1 r_1{}^n + c_2 r_2{}^n, \qquad (2.11.2)$$

where r_1 and r_2 are the roots of

$$r^2 + a_1 r + a_2 = 0, \qquad (2.11.3)$$

if r_1 and r_2 are distinct, we can readily obtain the relations corresponding to those given above for differential equations. The results will be stated as exercises for the reader.

EXERCISES

1. Obtain the explicit analytic solution of $u_{n+2} + a_1 u_{n+1} + a_2 u_n = 0$, $u_0 = b_1, u_1 = b_2$.
2. Show that a necessary and sufficient condition that all solutions of $u_{n+2} + a_1 u_{n+1} + a_2 u_n = 0$ approach zero is that the roots of $r^2 + a_1 r + a_2 = 0$ be less than one in absolute value.
3. What is a necessary and sufficient condition for this to hold?
4. If $u_{n+1} + a u_n = f_n$, $u_0 = c$, obtain an analytic expression for u_n.
5. Obtain the solution of $u_{n+2} + a_1 u_{n+2} + a_2 u_n = f_n$, $u_0 = b_1, u_1 = b_2$, using the method of generating functions (the discrete analogue of the Laplace transform), or the analog of the Lagrange variation-of-parameters technique.
6. If $u_{n+1} + a u_n = v_n$, $u_0 = c$, and $\lim_{n \to \infty} v_n$ exists, under what conditions does $\lim_{n \to \infty} u_n$ exist?
7. If

$$\lim_{N \to \infty} \left[a u_N + (1-a)(\sum_{k=1}^{N} u_n)/N \right]$$

exists with $0 < a < 1$, does $\lim_{N \to \infty} u_N$ exist?
8. If $u_n = c + a b^n$, $n \geq 1$, with $|b| < 1$, show that

$$c = u(\infty) = \frac{u_n u_{n+2} - u_{n+1}^2}{u_n + u_{n+2} - 2u_{n+1}}.$$

9. Obtain a corresponding expression for $u(\infty)$ if $u_n = c + a_1 b_1{}^n + a_2 b_2{}^n$, with $|b_1|, |b_2| < 1$.

10. Obtain the solution of the linear system $u_{n+1} = a_1 u_n + a_2 v_n$, $u_0 = c_1, v_{n+1} = b_1 u_n + b_2 v_n, v_0 = c_2$ by elimination of v_n.

11. If $u_{n+2} = a u_{n+1} + b u_n, u_0 = 1$, show that

$$\frac{u_{n+2}}{u_{n+1}} = a + \cfrac{b}{a + \cdots}.$$

(*Hint:* $u_{n+2}/u_{n+1} = a + b u_n/u_{n+1}$.)

12. Solve $r_{n+1} = (a r_n + b)/(c r_n + d), u_0 = a_1$, by setting $r_n = u_n/v_n$ where u_n, v_n are determined by a linear system of difference equations. (*Hint:* $u_{n+1}/v_{n+1} = (a u_n + b v_n)/(c u_n + d v_n)$.)

13. Obtain particular solutions of $u_{n+1} = a_1 u_n + a_2 v_n$, $v_{n+1} = b_1 u_n + b_2 v_n$ of the form $u_n = c_1 \lambda^n, v_n = c_2 \lambda^n$.

14. Establish the following summation by parts:

$a_0 b_0 + a_1 b_1 + \cdots + a_n b_n$

$\quad = B_0(a_0 - a_1) + B_1(a_1 - a_2) + \cdots + B_{n-1}(a_{n-1} - a_n) + a_n B_n,$

where $B_k = b_0 + b_1 + \cdots + b_k$.

15. Obtain the analog of the adjoint method for the linear difference equation $u_{n+2} + a_1 u_{n+1} + a_2 u_n = v_n$.

Miscellaneous Exercises

1. If $a(t) > 0$, show that no solution of $u'' - a(t)u = 0$ can have more than one zero. (*Hint:* Without loss of generality, let $u(t)$ be negative between the two points t_1 and t_2 where $u(t_1) = u(t_2) = 0$. Let t_3 be a point where $u(t)$ assumes a relative minimum value for $t_1 < t < t_2$. Show that this leads to a contradiction upon using the fact that $u'' - a(t)u = 0$.

2. Consider the Riccati equation $v' + v^2 - a(t) = 0$. Show that this equation has solutions which exist for all $t \geq 0$. (*Hint:* Recall the fact that $v = u'/u$.)

3. Consider the equation $\varepsilon u'' + (1 + \varepsilon)u' + u = 0$, $u(0) = 1$, $u'(0) = 0$, with $\varepsilon > 0$. Write the solution as $u(t, \varepsilon)$ to indicate the dependence on ε. Does $\lim_{\varepsilon \to 0} u(t, \varepsilon)$ exist? Does it satisfy the differential equation $u' + u = 0$? What is the initial condition?

4. Show that if $a(t) \geq 0$ then all solutions of $u'' + a(t)u' + u = 0$ remain bounded as $t \to \infty$. (*Hint:* $d/dt(u'^2 + u^2) = 2u'u'' + 2uu' = 2u'(-a(t)u' - u) + 2uu'$.)

5. By consideration of the equation $u'' + (2 + e^t)u' + u = 0$, show that the condition that $a(t) \geq a_0 > 0$ is not sufficient to ensure that all solutions approach zero as $t \to \infty$.

6. By means of a change of variable, show that $\varepsilon(t) \to 0$ as $t \to \infty$ cannot be a sufficient condition to ensure that all solutions of $\varepsilon(t)u'' + u' + u = 0$ approach a solution of $u' + u = 0$ as $t \to \infty$.

7. Consider the Sturm-Liouville equation $u'' + \lambda g(t)u = 0$, $u(0) = u(T) = 0$, where $g(t) \geq 0$. Let λ_1, λ_2 be two values for which a nontrivial solution exists and let u_1, u_2 be the associated functions. Then:

   (a) if $\lambda_1 \neq \lambda_2$ show that $\int_0^T g(t)u_1 u_2 \, dt = 0$, an orthogonality property. (*Hint:* Consider $u_1(u_2'' - \lambda_2 gu_2) - u_2(u_1'' - \lambda_1 gu_1) = 0$.);

   (b) hence, show that any such λ_i that exist must be real. (*Hint:* If λ is complex, $\bar{\lambda}$ is also an admissible value with the associated function $\bar{u}$.)

8. Consider the equation $u'' + \lambda u = 0$, $u(0) = 0$, $u(T) + au'(T) = 0$. Show that λ must satisfy the equation $\tan(\sqrt{\lambda}T) + a\sqrt{\lambda} = 0$ in order for a nontrivial solution to exist. Show that all roots of this equation are real and that there are an infinite number. Obtain the asymptotic expansion of the large roots. (For a systematic treatment of the conditions under which trigonometric expressions of the foregoing type appear, see

   R. Bellman and K. L. Cooke, *Differential-Difference Equations*, Academic Press, New York, 1963.)

9. If $u'' + u \geq 0$ in an interval length less than π and u is zero at the endpoints, must it preserve the same sign inside the interval?

10. Consider the equation $u'' + au' + bu = 0$, $u(0) = c_1$, $u'(0) = c_2$, and suppose that the roots of $r^2 + ar + b = 0$ have negative real parts. From the relations

$$\int_0^\infty u(u'' + au' + bu) \, dt = 0, \qquad \int_0^\infty u'(u'' + au' + bu) \, dt = 0,$$

plus integration by parts, obtain expressions for $\int_0^\infty u^2 \, dt$, $\int_0^\infty u'^2 \, dt$, as quadratic forms in c_1, c_2 without using the explicit form of u.

11. From the expression for $\int_0^\infty u^2 \, dt$ as a quadratic form in c_1 and c_2, derive a necessary and sufficient condition for the roots of $r^2 + ar + b = 0$ to have negative real parts.

12. Similarly, starting with $u^{(3)} + a_1 u^{(2)} + a_2 u^{(1)} + a_3 u = 0$, $u(0) = c_1$, $u'(0) = c_2$, $u''(0) = c_3$, derive expressions for $\int_0^\infty u^2 \, dt$, $\int_0^\infty u'^2 \, dt$, $\int_0^\infty u''^2 \, dt$, as quadratic forms in c_1, c_2, c_3, under the assumption that these integrals exist.

13. Hence, derive necessary and sufficient conditions that the roots of $r^3 + a_1 r^2 + a_2 r + a_3 = 0$ have negative real parts. (See

> A. Hurwitz, " Uber die Bedingungen unter welchen eine Gleichung nur Wurzeln mit negativen reellen Teilen besitzt," *Math. Ann.*, **46**, 1895 (Werke, **2**, pp. 533–545),
>
> H. Cremer and F. H. Effertz, " Uber die algebraische Kriterien fur die Stabilitat von Regelungssystemen," *Math. Ann.*, **137**, 1959, pp. 328–350.)

14. Consider the Green's function $k(t, t_1)$ associated with $u'' = f$, $u(0) = u(T) = 0$. Show that

$$\int_0^T k(t, t_1) k(t_1, t_2) \, dt_1 = k(t, t_2),$$

for $0 \le t, t_1, t_2 \le T$.

15. Establish this result for the Green's function of the second-order linear differential equation $u'' + a(t)u = f$, $u(0) = u(T) = 0$.

16. Show that the Green's function of the foregoing equation can be characterized by the condition that $k(t, s)$ is a solution of the linear differential equation in $(0, s)$ and $(s, 1)$, with a discontinuity derivative at $t = s$, namely

$$k'(s - 0, s) - k'(s + 0, s) = 1$$

for $0 < s < 1$.

BIBLIOGRAPHY AND COMMENTS

2.1. A more leisurely account of the material covered in this chapter may be found in

R. Bellman, *Modern Elementary Differential Equations*, Addison-Wesley, Reading, Massachusetts, 1967.

For intermediate and advanced results, see

E. Coddington and N. Levinson, *Theory of Ordinary Differential Equations*, McGraw-Hill, New York, 1955.

E. L. Ince, *Ordinary Differential Equations*, Dover, New York, 1944.

3

STABILITY AND CONTROL

3.1 Introduction

Our aim is to present some of the background in classical stability that motivates the continuing concern with control theory that one sees in the engineering and mathematical journals.

We begin with the observation that one of the basic objectives of science is to predict the future. This clairvoyance may range from that of the date of the next eclipse, earthquake, or recession, to that of the failure of a structure such as a building or bridge under sustained vibration.

In this effort to foretell, the crystal ball of science is the classical descriptive approach, aided and amplified by mathematical ingenuity and the electronic computer. To illustrate what we mean, consider a system S whose state at any time t can be described by a scalar function, $u(t)$, a "displacement," and its derivative, $u'(t)$, a "velocity."

"Acceleration" and higher-order rates of change can be determined from these quantities by means of a relation that takes the form of a differential equation

$$u'' = g(u, u'), \tag{3.1.1}$$

with assigned initial conditions

$$u(0) = c_1, \qquad u'(0) = c_2. \tag{3.1.2}$$

28

In general, systems cannot be described in such a compact fashion. Many further state variables may be required and more complicated functional relations requiring the past history of the system may be necessary.

In a number of cases, however, a closure relation of the foregoing type is a convenient initial approximation, an important starting point for more realistic studies.

Under reasonable assumptions concerning the function $g(u, u')$, the solution of (3.1.1) exists over some initial time $[0, t_0]$ and is uniquely determined by the initial state. Thus, a knowledge of the current state enables us to predict the behavior of the system over some future time interval. For this to be an operational technique, we must possess analytic and computational methods that enable us to extract this information in less time than that required to wait for the anticipated event. This is a major consideration in mathematical theories of weather prediction, chemotherapy, and respiratory control.

Let us now discuss a simple example. We may be interested in the period of oscillation of a pendulum clock; see Fig. 3.1. Allowing for a

Figure 3.1

small damping effect due to friction at 0, taken to be proportional to angular velocity, taking account of the influence of gravity, and making a number of plausible idealizations, we obtain an equation of the type exhibited in (3.1.1),

$$u'' + au' + \sin u = 0, \qquad (3.1.3)$$

where we have normalized various physical units. Here $a > 0$, corresponding to a frictional effect. We are now in a position to obtain the period of damped oscillation, the time required to come to rest, and so on.

Much of classical analysis is devoted to the answering of questions of this nature, the systematic study of the qualitative and quantitative properties of solutions of differential equations. For example, we may be interested in the existence of periodic solutions, in the asymptotic behavior of the solution as t becomes infinite, or in various analytic approximations involving power series or the solution of differential equations of lower degree.

This type of investigation has been fruitfully extended by considering more general types of functional equations arising in the study of other physical systems, partial differential equations such as

$$u_t = u_{xx} + g(u, u_x), \tag{3.1.4}$$

and differential-difference equations such as

$$u'' = g(u, u'; u(t - \tau), u'(t - \tau)). \tag{3.1.5}$$

Furthermore, by considering stochastic processes of different kinds, many more intriguing categories of equations are engendered.

In this volume, we will restrict our attention to finite-dimensional processes of deterministic type, and even here we will consider only particularly simple kinds of equations. We try as carefully as we can to avoid a confusion of purely analytic difficulties due to the nature of the equation with conceptual complexities inherent in a theory of control. It is for this reason that we consider the one-dimensional control process first for orientation purposes.

3.2. Stability

One of the most interesting and important qualitative properties of a solution is that of stability. By this term, we mean intuitively the ability of the solution of the equation to preserve certain structural features under various types of changes in the values of the initial conditions, or the form of the equation, or both.

To illustrate one aspect of this fundamental concept of both mathematics and science, consider the equation

$$w'' + aw' + \sin w = g(t), \tag{3.2.1}$$

which we interpret as describing the pendulum of Section 3.1, subject to some external forces. If $g(t) = 0$, we know that $w(t) = 0$ is a solution. This corresponds to a position of equilibrium.

If we assign the initial conditions

$$w(0) = w'(0) = 0, \qquad (3.2.2)$$

and impose the condition that the external disturbance is small at any particular time, if, for example, for $t \geq 0$ we have

$$|g(t)| \leq \varepsilon, \qquad (3.2.3)$$

can we assert that the deviation from equilibrium of the pendulum remains small? Is it true that

$$|w| + |w'| \leq \varepsilon_1, \qquad (3.2.4)$$

where ε_1 is some function of ε, for example, $\varepsilon_1 = k\varepsilon$? If this is so, we say that $w = 0$, $w' = 0$ is a position of stable equilibrium.

Can small forcing effects build up over time to produce a major effect? Are there nonlinear resonances whereby small external oscillations can induce a significant oscillation inside the system?

These are questions that are essential to ask in view of the fact that we know that no mathematical equation describes a physical system exactly. Certain idealizations are always made in the construction of a mathematical model as, for example, in the neglect of certain state variables, the omission of certain interactions that simplify cause and effect relations, the simplification of the geometric structure, the assumption of instantaneous transmission of certain effects, the overlooking of various possible changes in the environment, and so on, and so on. To compensate for the effect of this initial smoothing of the scientific path, we introduce an all-purpose forcing term $g(t)$, representing the resultant influence of variables and processes that have been ignored. If we then wish to use the solution of the modified equation to predict the behavior of the original system with some degree of confidence, we cannot avoid consideration of the foregoing questions.

Alternatively, consider the case where we assume that some chance phenomenon has dislodged the system from its equilibrium position. In place of the equation

$$u'' + au' + \sin u = 0, \qquad u(0) = u'(0) = 0, \qquad (3.2.5)$$

with the unique solution $u(t) = 0$, for $t \geq 0$, we have the equation

$$w'' + aw' + \sin w = 0, \qquad w(0) = \varepsilon_1, \qquad w'(0) = \varepsilon_2, \qquad (3.2.6)$$

where $|\varepsilon_1|$ and $|\varepsilon_2|$ are both small.

Two questions are now pertinent:

(a)   Is it true that $|w'| + |w| \leq k_1(|\varepsilon_1| + |\varepsilon_2|)$ for $t \geq 0$
      for some constant k_1?   (3.2.7)
(b)   Is it true that $\lim_{t \to \infty}(|w'| + |w|) = 0$?

This is to say that we want to know whether or not a small initial disturbance remains small, and, if so, whether it dies out completely over time.

The foregoing problem corresponds to giving the pendulum a slight push when it is in the downward position (see Fig. 3.2). There is, however, another equilibrium position (see Fig. 3.3).

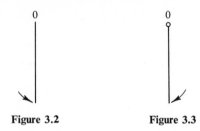

Figure 3.2   Figure 3.3

The initial conditions corresponding to this position are

$$u(0) = \pi, \qquad u'(0) = 0. \qquad (3.2.8)$$

The corresponding solution is $u(t) = \pi$, $t \geq 0$. We are interested in the answers to the same questions as before.

Without any detailed mathematical analysis (which we carry out below anyway), we can predict what will happen. With a considerable amount of experience in the use of the foregoing mathematical model of the motion of a pendulum as a basis for the explanation of the behavior of an actual pendulum, we know that in the first case the position of equilibrium is stable, while in the second case it is not.

The situation is not at all as obvious in connection with the study of the mathematical investigation of the equations describing the motion of the moon and the earth around the sun when both are taken as point particles. We are reasonably sure that the solar system is stable as far as the perturbation due to a stray comet or spaceship is concerned, but we have no proof. This type of question occupied a great deal of the attention of the mathematical world of the eighteenth and nineteenth centuries and produced a number of important mathematical theories. These have long since escaped the confines of the original problem (an interesting instability in itself), and have been applied both to answer many questions of more pressing and immediate concern and to create new groves of academe.

3.3. Numerical Solution and Stability

A third type of stability question arises very naturally in connection with the numerical solution of differential equations and of functional equations in general. To illustrate this, consider the first-order differential equation

$$u' = g(u), \qquad u(0) = c. \tag{3.3.1}$$

To obtain a numerical solution accurate to a specified number of significant figures we can consider the difference equation

$$\frac{v(t + \Delta) - v(t)}{\Delta} = g(v(t)), \qquad v(0) = c, \tag{3.3.2}$$

where t now assumes the values $0, \Delta, \ldots$. Actually, this is a very poor difference approximation to use, but this is beside the point here. The question is whether or not the sequence $\{v(k\Delta)\}$, $k = 0, 1, 2, \ldots$, calculated using (3.3.2) is a reasonable approximation to the sequence $\{u(k\Delta)\}$ obtained from (3.3.1).

This may be viewed as a stability problem in the following fashion. From (3.3.1), we have upon integrating between t and $t + \Delta$,

$$\frac{u(t + \Delta) - u(t)}{\Delta} = \frac{1}{\Delta} \int_t^{t+\Delta} g(u) \, dt = g(u(t)) + \delta(t, \Delta), \tag{3.3.3}$$

where $\delta(t, \Delta)$ is small if Δ is small. Hence, we see that the question of the degree of approximation of $v(k\Delta)$ to $u(k\Delta)$ is one concerning the stability of the solution of the difference equation in (3.3.2). It can be treated by the Poincaré-Lyapunov theory we discuss briefly below.

We see that the use of a quadrature technique produces one error in (3.3.3). The calculation of $g(u(t))$ in any arithmetic fashion where only K significant figures are retained produces another error, "round-off" error.

Do these errors mount up as the calculation proceeds and eventually overwhelm the desired solution? This is the basic question of the field of numerical analysis. Although we shall not discuss it here, it is important to note that numerical calculation can profitably be considered to be a control process in which the final error is to lie within prescribed bounds. This means, in particular, that we can either use a fixed algorithm or a sequential approximation technique based upon feedback control ideas.

3.4. Perturbation Procedures

Let us return to the equation of the pendulum,

$$u'' + au' + \sin u = 0, \qquad u(0) = \varepsilon_1, \qquad u'(0) = \varepsilon_2, \qquad (3.4.1)$$

where ε_1 and ε_2 are small quantities. Since it is not easy to examine the solution directly in order to answer the questions of (3.2.7), even in the case $a = 0$ where the solution can be obtained in terms of elliptic functions, we use an approach to the persistence of equilibrium states systematically exploited by eighteenth- and nineteenth-century analysts.

Since we are interested, to begin with, in what happens when the system is disturbed by small effects, let us use the linear approximation

$$\sin u \cong u \qquad (3.4.2)$$

to simplify (3.4.1). The new equation

$$w'' + aw' + w = 0, \qquad w(0) = \varepsilon_1, \qquad w'(0) = \varepsilon_2, \qquad (3.4.3)$$

is a linear equation with constant coefficients possessing a solution that can be easily obtained in explicit analytic form. As we have already indicated, the solution is

$$w = b_1 e^{\lambda_1 t} + b_2 e^{\lambda_2 t}, \qquad (3.4.4)$$

where λ_1, λ_2 are the roots of the characteristic equation

$$\lambda^2 + a\lambda + 1 = 0 \qquad (3.4.5)$$

(distinct if a is small), and b_1, b_2 are determined by the equations

$$b_1 + b_2 = \varepsilon_1,$$
$$\lambda_1 b_1 + \lambda_2 b_2 = \varepsilon_2. \tag{3.4.6}$$

The roots are

$$\lambda_1, \lambda_2 = \frac{-a \pm (a^2 - 4)^{1/2}}{2}, \tag{3.4.7}$$

conjugate complex with negative real part if $0 < a < 2$. If $a > 2$, the "over-damped" case, both are negative. The quantities b_1, b_2 are linear combinations of $\varepsilon_1, \varepsilon_2$. We see that all solutions approach zero as $t \to \infty$.

It is thus plausible that small disturbances of the position of the pendulum in Fig. 3.2 are ultimately damped out and that in the course of this damped oscillatory behavior, the pendulum never deviates too far from the equilibrium position in state space. A precise statement will be found in Section 3.5.

Consider, however, the situation in the case where the initial position is that of Fig. 3.3. If we write $u = \pi + w$, we obtain the equation

$$w'' + aw' + \sin(\pi + w) = 0, \qquad w(0) = \varepsilon_3, \qquad w'(0) = \varepsilon_4. \tag{3.4.8}$$

Writing

$$\sin(\pi + w) = -\sin w \cong -w, \tag{3.4.9}$$

we obtain the appropriate equation

$$w'' + aw' - w = 0, \qquad w(0) = \varepsilon_3, \qquad w'(0) = \varepsilon_4. \tag{3.4.10}$$

The characteristic roots are now

$$\lambda_3, \lambda_4 = \frac{-a \pm (a^2 + 4)^{1/2}}{2}, \tag{3.4.11}$$

one of which is positive and one of which is negative, regardless of the value of a.

For general values of ε_3 and ε_4, $w(t)$ increases without bound as $t \to \infty$. Hence, we suspect that the equilibrium position of Fig. 3.3 is unstable.

3.5. A Fundamental Stability Theorem

The foregoing procedures can be validated on the basis of a general stability theorem that can be used to study the behavior of far more complicated systems. A tremendous amount of effort has been devoted to this area of the theory of differential equations, and, as current journals will attest, this is by no means a closed chapter of analysis.

A two-dimensional version of a fundamental theorem of Poincaré and Lyapunov is the following theorem.

Theorem. *Consider the equation*

$$u'' + au' + bu = g(u, u'), \qquad u(0) = c_1, \qquad u'(0) = c_2, \qquad (3.5.1)$$

where $g(u, u')$ satisfies a Lipschitz condition,† to ensure existence and uniqueness, and where, in addition,

(a) *The roots of $\lambda^2 + a\lambda + b = 0$ have negative real parts.*
(b) $|g(u, v)|/(|u| + |v|) \to 0$ *as* $|u| + |v| \to 0.$ (3.5.2)
(c) $|c_1| + |c_2| \leq \delta$ *where δ depends upon the function g.*

Then the solution of (3.5.1) approaches zero as $t \to \infty$.

If a characteristic root has positive real part, the equilibrium solution $u = 0$ is not stable.

This result assures us that the stability of the equilibrium position can be determined by studying the properties of the equation of linear approximation,

$$w'' + aw' + bw = 0, \qquad (3.5.3)$$

in the cases where no characteristic roots with zero real part occur.

In the case where $a = 0$, the situation is far more complex, due to the fact that the nonlinear terms now play an essential role.

---

† This will be the case if $g(u, u')$ possesses bounded partial derivatives with respect to both arguments in the neighborhood of $u = c_1$, $u' = c_2$.

3.6. Stability by Design

The importance of the Poincaré-Lyapunov theorem resides in the fact that the design of a system that is to be stable under small perturbations is reduced to an algebraic problem, namely, the determination of the location of the roots of the characteristic polynomial associated with the differential equation describing its behavior.

In the two-dimensional case, where the characteristic polynomial is $\lambda^2 + a\lambda + b$, it requires no effort to obtain necessary and sufficient conditions. In the case of multidimensional systems where the characteristic polynomial is $\lambda^N + a_1\lambda^{N-1} + \cdots + a_N$, a considerable amount of ingenuity is required. Many useful criteria now exist, notably those of Hurwitz and Routh.

3.7. Stability by Control

In a number of important situations the system has already been constructed, which means that we no longer possess any choice of the components, or, equivalently, of the parameters a and b of Section 3.5. In order to eliminate some undesired behavior of the system, or to maintain it in some desired state, we must introduce some additional forces.

Schematically, see Fig. 3.4. Here $v(t)$ represents the monitoring action.

Thus, for example, we may wish to maintain the pendulum in the unstable position shown in Fig. 3.3, by external influences applied at various time intervals. Sometimes, we merely wish to reinforce stability.

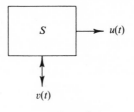

Figure 3.4

It may be desirable for the system to return to the desired equilibrium position as quickly as possible, or we may wish to eliminate minute but persistent oscillations that are either annoying to humans, or that tend to produce structural fatigue.

Consider the equation

$$u'' + au' + u = 0, \tag{3.7.1}$$

where a is positive, but so small that the solution is oscillatory for all practical purposes. One way to damp out the oscillation is to introduce an external force $v(t)$ designed to counteract the intrinsic behavior of the system. Suppose that the new equation is

$$u'' + au' + u = v(t). \tag{3.7.2}$$

How should this function $v(t)$ be chosen? There is no difficulty in choosing $v(t)$ to dispose of any oscillation corresponding to a particular initial disturbance, $u(0) = c_1, u'(0) = c_2$.

This choice of $v(t)$ may not be satisfactory for a different set of values of c_1 and c_2. We want to choose a forcing function that will ensure that the system returns to equilibrium position rapidly regardless of the nature of the small perturbation, or the time at which it occurs.

One way to accomplish this is to take $v(t)$ not to depend upon the time t, but rather upon the state of the system, u and u'. Thus, we write $v = g(u, u')$, and consider the new equation

$$u'' + au' + u = g(u, u'). \tag{3.7.3}$$

This is an application of the general concept of *feedback control.* Schematically, see Fig. 3.5. By this diagram, we wish to emphasize that the controlling action that is exerted at any time depends upon the state of the system at that time, in this case the deviation from equilibrium as indicated by the values u and u'. This is a common-sense point of view.

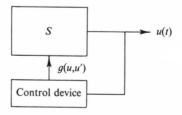

Figure 3.5

3.8. Proportional Control

It is reasonable to suspect that the simplest type of feedback control is that where $g(u, u')$ is a linear function of u and u'. This is actually not the case, but it would take us too far afield to discuss the matter in any meaningful fashion. The point is that the term "simple" has to be evaluated in both its engineering and mathematical contexts.

For the purposes of exposition, let us take

$$g(u, u') = f_1 u + f_2 u', \tag{3.8.1}$$

where f_1 and f_2 are constants at our disposal, and consider the equation

$$u'' + au' + u = f_1 u + f_2 u'. \tag{3.8.2}$$

Our aim is to determine f_1 and f_2 so that the solution of (3.8.2) approaches the equilibrium position, $u = u' = 0$, rapidly.

To begin with, suppose that we want to use a feedback control system based only upon a knowledge of $u(t)$, the displacement. Take then $f_2 = 0$, with f_1 remaining to be chosen. The linear differential equation is

$$u'' + au' + (1 - f_1)u = 0, \tag{3.8.3}$$

with the characteristic equation

$$\lambda^2 + a\lambda + (1 - f_1) = 0. \tag{3.8.4}$$

The characteristic roots are

$$\lambda_1, \lambda_2 = \frac{-a \pm (a^2 - 4(1 - f_1))^{1/2}}{2}. \tag{3.8.5}$$

If $a^2 < 4(1 - f_1)$, the roots are complex with real part $-a/2$. If $a^2 > 4(1 - f_1)$, the roots are real, with one greater than $-a/2$ and one less. This means an increase in the time required to damp out an arbitrary disturbance.

We see then that no linear feedback control based solely upon knowledge of the displacement will increase the rate of damping.

Suppose that we try proportional control based upon an observation of the rate of displacement, u'. The corresponding equation is

$$u'' + au' + u = f_2 u'. \tag{3.8.6}$$

The characteristic equation is

$$\lambda^2 + (-f_2 + a)\lambda + 1 = 0. \tag{3.8.7}$$

By taking $-f_2$ a large positive constant, we can make the damping rate as large as we wish. Presumably, this disposes of the problem.

But recall that the use of a linear equation in the study of stability is predicated upon the hypothesis that the displacement and velocities are small; for example, we made the approximation $\sin u \cong u$. Suppose we take $a - f_2 = k$ to be a large positive constant. Consider the differential equation

$$u'' + ku' + u = 0. \tag{3.8.8}$$

The two principal solutions are approximately

$$u_1 = e^{-kt/2} \cos \frac{kt}{2}, \qquad u_2 = \left(2e^{-kt/2} \sin \frac{kt}{2}\right)\Big/ k. \tag{3.8.9}$$

The function u_1 is zero at $t = \pi/k$, but its derivative is

$$u_1'\left(\frac{\pi}{k}\right) = -ke^{-\pi/2}. \tag{3.8.10}$$

Hence, $|u_1'|$ can be quite large if k is large, which violates our fundamental assumption of small perturbations.

3.9. Discussion

Rudimentary as the preceding discussion has been, several interesting consequences can be deduced. In the first place, we note that the effectiveness of feedback control depends strongly upon the type of information available concerning the system. We will leave the implications of this basic idea aside until the third volume.

For the present, the essential point we wish to make is that arbitrarily effective control is impossible. The effect of control on one aspect of the behavior of a system must be balanced against the reactions in another part of the system or on another type of behavior of the system. In the foregoing case, if we attempt to return the system to equilibrium too quickly, we generate a velocity that can be far more disastrous than the effect we are trying to counteract.

Another way of putting this is to say that we must weigh the cost of undesirable performance of the system against the cost of control. This is one of the basic facts of control theory: nothing for nothing.

How do we evaluate and compare the various consequences? There are many different approaches, none uniformly satisfactory. Each control process must be examined on its own merits. In what follows, we will consider one general approach that has certain advantages that will be indicated.

3.10. Analytic Formulation

Consider the type of control process discussed above. We have the equation

$$u'' + au' + u = g(u, u'), \tag{3.10.1}$$

and we would like to choose the function $g(u, u')$, the feedback control law, in such a way that u and u' approach zero reasonably rapidly without violating the condition that $|u|$ and $|u'|$ must not be too large. This is clearly not a precise analytic formulation.

It can be made precise in several ways at the expense of introducing analytic difficulties of a higher order of difficulty than those we care to encounter at this stage of the proceedings.

Let us then compromise our original aims, the constant lot of the research scientist, and ask for a control law that makes u and u' small on the average. For example, let us try to determine the function g so that the quantity

$$\int_0^T (u^2 + u'^2)\, dt \tag{3.10.2}$$

is small. This formulation possesses certain merits, notably analytic simplicity.

We have, however, omitted to impose any cost of control. Different types of control laws require different amounts of effort to implement. How should we estimate this type of cost? One way is to use some average cost, such as

$$\int_0^T g(u, u')^2\, dt. \tag{3.10.3}$$

If we add the two costs together, we have a total cost of

$$\int_0^T [u^2 + u'^2 + g(u, u')^2]\, dt. \qquad (3.10.4)$$

How do we choose the function g so as to miminize this total cost?

We can now proceed to study questions of this nature in two ways, directly using dynamic programming, or by means of the calculus of variations using a simple artifice. Dynamic programming will be employed in Chapter 5 and the calculus of variations in Chapter 4. The order of presentation is a consequence of reasons that will become clear as we proceed.

Let us prepare the way for the calculus of variations by indicating how we can simplify the foregoing formulation. The desired function $g(u, u')$ is a function of t, through the dependence of u and u' upon t. Why not then write

$$u'' + au' + u = v(t), \qquad (3.10.5)$$

and determine $v(t)$ by the condition that it minimizes the functional

$$\int_0^T [u^2 + u'^2 + v^2]\, dt\ ? \qquad (3.10.6)$$

It is essential to include the initial conditions $u(0) = c_1, u'(0) = c_2$. Hopefully, once $v(t)$ has been determined, we can interpret $v(t)$ as a function of the state $u(t)$ and $u'(t)$, and thus obtain the desired feedback law. This simplification possesses the overwhelming merit of allowing us to deal with linear differential equations such as (3.10.5) rather than nonlinear equations such as (3.10.1), in the description of the behavior of the system over time.

3.11. One-Dimensional Systems

It turns out that it is wise initially to consider the simplest system, one described by a single state variable, a displacement $u(t)$. Once we have carried through the analysis in this case, we can turn to the general case.

We consider then a system S described by a state variable $u(t)$ satisfying the equation

$$\frac{du}{dt} = g(u, v), \qquad u(0) = c, \tag{3.11.1}$$

where the control variable $v(t)$ is to be chosen so as to minimize the functional

$$J(u, v) = \int_0^T h(u, v)\, dt, \tag{3.11.2}$$

where $h(u, v)$ is a prescribed function. This is still too complex a problem for a suitable introduction to the analytic aspects of control theory. Consequently, we use a linear approximation in (3.11.1),

$$\frac{du}{dt} = au + v, \qquad u(0) = c, \tag{3.11.3}$$

and a quadratic approximation in (3.11.2).

$$J(u, v) = \int_0^T [u^2 + v^2]\, dt. \tag{3.11.4}$$

Finally, we begin by taking $a = 0$. We thus have the specific problem of minimizing the functional

$$J(u) = \int_0^T [u^2 + u'^2]\, dt, \tag{3.11.5}$$

where $u(0) = c$.

This will be our starting point in the following chapter. Once the basic ideas of the solution of this problem are made clear, it is not difficult to consider classes of more general variational problems. However, it is not to be anticipated that any all-purpose theory capable of handling completely general variational problems will ever be developed. The reason is quite simple. We face all the difficulties of treating general nonlinear differential equations, plus difficulties of a higher order, which we will indicate subsequently.

What we can hope to do is to develop a toolchest of techniques that can be used singly, in unison, and together with an electronic computer to handle broader and broader classes of problems.

Miscellaneous Exercises

1. Consider the first-order nonlinear differential equation $u' = -u + u^2$, $u(0) = c$. Prove by examination of the explicit solution that u exists for $t > 0$ provided that $|c|$ is sufficiently small. Show that if $|c|$ is sufficiently small, then $|u| \leq 2|c|$ for $t > 0$.

2. Consider the first-order equation $u' = -u + g(u)$, $u(0) = c$, where $|g(u)| \leq k_1|u|^2$ for $|u|$ small. Show that u exists for $t \geq 0$ for $|c|$ small, and indeed that $|u| \leq 2|c|$, provided that $|c|$ is small, in the following steps:

    (a) u satisfies the integral equation

$$u = e^{-t}c + \int_0^t e^{-(t-t_1)}g(u)\,dt_1.$$

    (b) $|u| < 2|c|$ in some initial interval $[0, a]$.
    (c) Then, if $|u| = 2|c|$ for $t = b$, we have

$$2|c| = |u| \leq e^{-b}|c| + \int_0^b e^{-(b-t_1)}|g(u)|\,dt_1$$

$$\leq |c| + 4k_1|c|^2 \int_0^b e^{-(b-t_1)}\,dt_1$$

$$< |c| + 4k_1|c|^2 \int_0^\infty e^{-t_1}\,dt_1,$$

    a contradiction if $|c|$ is sufficiently small.

3. Extend the argument to show that $|u| < 2|c|e^{-t}$ for $t \geq 0$, if $|c|$ is small.

4. Extend this method of proof, step-by-step, to cover the case where u satisfies the second-order equation $u'' + au' + bu = g(u, u')$, $u(0) = c_1$, $u'(0) = c_2$, under the following assumptions:

    (a) The roots of $r^2 + ar + b = 0$ have negative real parts.
    (b) $|g(u, u')| \leq k_1(|u| + |u'|)^2$ if $|u| + |u'|$ is sufficiently small.
    (c) $|c_1| + |c_2|$ is small.

    Show that the integral equation for u has the form

$$u = u_0 + \int_0^t k(t - t_1)g(u, u')\,dt_1.$$

Show that $|u| \le k_2(|c_1| + |c_2|)e^{-r_1 t}$ for a suitable r_1 and k_2. (*Hint:* Consider the equations for both u and u'.)

5. Consider the equation $u' = -u + g(u)$, $u(0) = c$, under the foregoing hypotheses. To show that $|u|$ remains bounded, consider the function u^2. We have

$$\frac{d}{dt}(u^2) = 2uu' = 2u(-u + g(u)) = -2u^2 + 2ug(u).$$

By assumption, $|g(u)|/|u|$ is small for small $|u|$. Hence, if $|c|$ is small, we have $d/dt(u^2) < 0$. Thus, u^2 decreases as t increases, and thus $|u|$ is uniformly bounded for $t \ge 0$.

6. Extend the foregoing argument to show that $|u| \le e^{-r_1 t}|c|$ for a suitable r_1. (*Hint:* We have $d/dt(u^2) < -au^2$ for small t and $|c|$.)

7. Consider the equation $u'' + au' + u = 0$, with $a > 0$. Show that $u^2 + u'^2$ is monotonically decreasing for $t \ge 0$. (*Hint:* $d/dt(u^2 + u'^2) = 2uu' + 2u'(-au' - u) = -2au'^2$.)

8. Show that if $u'' + au' + bu = 0$, a, $b > 0$, there is a corresponding quadratic expression in u and u' that is monotone decreasing.

9. Consider the quadratic expression $a_1 u_1^2 + 2a_2 u_1 u_2 + a_3 u_2^2$. If $a_1 > 0$, $a_2^2 < a_1 a_2$, then there exist constants b_1, $b_2 > 0$ such that $b_1(u_1^2 + u_2^2) \le a_1 u_1^2 + 2a_2 u_1 u_2 + a_3 u_2^2 \le b_2(u_1^2 + u_2^2)$.

10. Let u satisfy the equation of Exercise 7. Show that constants a_1, a_2 can be found such that $u^2 + a_1 uu' + a_2 u'^2$ is positive for all non-trivial values of u and u' and such that

$$\frac{d}{dt}(u^2 + a_1 uu' + a_2 u'^2) \le -a_3(u^2 + u'^2)$$

$$\le -a_4(u^2 + a_1 uu' + a_2 u'^2),$$

where a_3, $a_4 > 0$.

11. Hence, show that u, $u' \to 0$ as $t \to \infty$ and indeed that $u^2 + u'^2 \le a_5 e^{-a_6 t}$ for some positive constants a_5, a_6.

12. Using the foregoing techniques, show that $u'' + au' + u = g(u, u')$, $u(0) = c_1$, $u'(0) = c_2$, where the conditions of Exercise 4 are satisfied, then $u^2 + u'^2 \le a_7 e^{-a_8 t}$ for some positive constants a_7, a_8. (This is a particular application of the powerful "second method" of Lyapunov.)

13. Using the same techniques as in Exercise 4, study the behavior of the solution of $u'' - u = g(u, u')$, $u(0) = c$, $u(T) = 0$, as $T \to \infty$,

under the assumptions (a) $|c|$ is small; (b) $|g(u, u')| \leq k_1(u^2 + u'^2)$ for $|u|$, $|u'|$ small. (See

R. Bellman, "On Analogues of Poincaré-Lyapunov Theory for Multipoint Boundary-Value Problems—I," *J. Math. Anal. Appl.*, **14**, 1966, pp. 522–526.)

14. Consider the Riccati equation $u' = -u + u^2 + g(t)$, $u(0) = c$. Show that $|u|$ is bounded for $t \geq 0$ if $|c|$ and $\max_{t>0} |g(t)|$ are sufficiently small.

15. Consider the equation $u'' - u = 0$, $u(0) = 1$, $u(T) = e^{-T}$ with the solution $u = e^{-t}$. Compare this with solution of $v'' - v = \varepsilon_1$, $v(0) = 1 + \varepsilon_2$, $v(T) = e^{-T}$. Is $\max_{0 \leq t \leq T} |u - v|$ small if $|\varepsilon_1|$ and $|\varepsilon_2|$ are small?

16. The equation $u'' - u = 0$, $u(0) = 1$, $u(\infty) = 0$ has the unique solution $u = e^{-t}$. Does $v'' - v = \varepsilon_1$, $v(0) = 1 + \varepsilon$, $v(\infty) = 0$ possess a solution for arbitrary values of ε_1 and ε_2?

BIBLIOGRAPHY AND COMMENTS

There are many books which contain extensive discussions of nonlinear difference equations and stability theory. Let us cite

R. Bellman, *Stability Theory of Differential Equations*, McGraw-Hill, New York, 1953.

R. Bellman, I. Glicksberg, and O. Gross, *Some Aspects of the Mathematical Theory of Control Processes*, The RAND Corporation, **R-313**, 1958.

L. Cesari, *Asymptotic Behavior and Stability Problems in Ordinary Differential Equations*, Academic Press, New York, 1963.

N. Minorsky, *Nonlinear Oscillations*, Van Nostrand, Princeton, New Jersey, 1962.

A useful survey of recent work is

R. W. Brockett, "The Status of Stability Theory for Deterministic Systems," *AIAA*, **3**, 1965, pp. 596–606.

For historical background, see

R. Bellman and R. Kalaba, *Mathematical Trends in Control Theory*, Dover, New York, 1964.

For the stability of differential-difference equations, see

R. Bellman and K. L. Cooke, *Differential-Difference Equations*, Academic Press, New York, 1963.

3.1. For an indication of the novel types of functional equations that arise in chemotherapy and respiratory control, see

R. Bellman, J. Jacquez, and R. Kalaba, "Some Mathematical Aspects of Chemotherapy—I: One-Organ Models," *Bull. Math. Biophys.*, **22**, 1960, pp. 181–198.

R. Bellman, J. Jacquez, and R. Kalaba, "Some Mathematical Aspects of Chemotherapy—II: The Distribution of a Drug in the Body," *Bull. Math. Biophys.* **22**, 1960, pp. 309–322.

R. Bellman, J. Jacquez, and R. Kalaba, "Mathematical Models of Chemotherapy," *Proc. Fourth Berkeley Symposium on Mathematical Statistics and Probability*, Univ. California Press, Berkeley, California, **I**, 1961, pp. 37–48.

J. Buell, F. S. Grodins, and A. J. Bart, "Mathematical Analysis and Digital Simulation of the Respiratory Control Systems," *J. Appl. Physiol.*, Feb., 1967, pp. 260–276.

3.2. Detailed discussions of equations of this type plus further references will be found in the books by Cesari and Minorsky cited above. There has been virtually no systematic discussion of techniques of mathematical formulation of physical processes.

3.3. For a detailed discussion of this material, see

R. Bellman and J. M. Richardson, *Introduction to Methods of Nonlinear Analysis*, to appear.

3.4. For a brief introduction to perturbation techniques, see

R. Bellman, *Perturbation Techniques in Mathematics, Physics and Engineering*, Holt, New York, 1964.

3.5. Proofs of this basic theorem, together with extensions and numerous references, will be found in the books by Cesari and Bellman cited above.

3.6. Analysis of this type was initiated by Maxwell and Vishnegardskii, stimulated by feedback control devices such as the governor on the Watt steam engine. The path from this to Nyquist diagrams in the *s*-plane is an interesting one. See the Bellman and Kalaba book cited above.

3.8. The difficulty with linear control, and indeed with feedback control in general, is that accurate sensing devices are needed to measure the state of the system and often costly devices are required to supply the forcing term $g(u, u')$. One way of overcoming the engineering complications is to employ "bang-bang" control where $g(u, u')$ can assume only the values ± 1. The control is either on or off.

There are many intriguing analytic aspects to the determination of optimal control, some of which will be discussed in Volume IV. The problem was first discussed in

D. W. Bushaw, *Differential Equations with a Discontinuous Forcing Term*, Ph.D. Thesis, Department of Mathematics, Princeton Univ., Princeton, New Jersey, 1952.

See also

D. W. Bushaw, "Optimal Discontinuous Forcing Terms," *Contributions to the Theory of Nonlinear Oscillations*, *Vol. IV*, Princeton Univ. Press, Princeton, New Jersey, 1958.

For a treatment based upon function space techniques, see

R. Bellman, I. Glicksberg, and O. Gross, "On the 'bang-bang' Control Problem,"
     Quart. Appl. Math., **14**, 1956, pp. 11–18.
J. P. LaSalle, "The Time-Optimal Control Problem," *Contributions to Differential Equations*, Vol. V, Princeton Univ. Press, Princeton, New Jersey, 1960, pp. 1–24.

3.10. Observe that the direct feedback control formulation involving $g(u, u')$ requires that we be able to observe the system accurately in order to obtain the values of u and u'. On the other hand, the formulation in terms of $v(t)$ requires that we know the time quite accurately.

As soon as we begin the discussion of control processes involving quantum mechanical and relativistic effects, both of these requirements assume important roles.

3.11. Basically, our hope is that we can begin with the linear equation and quadratic functional of (3.10.3) and (3.10.4) and use successive approximations to treat the general equation of (1) and the general functional of (3.10.2). This will be discussed in Volume II.

4

CONTINUOUS VARIATIONAL PROCESSES;
CALCULUS OF VARIATIONS

4.1. Introduction

In this chapter, we wish to study the problem of minimizing the functional

$$J(u) = \int_0^T (u'^2 + u^2)\, dt \qquad (4.1.1)$$

using the methods of the calculus of variations. We will first consider the case where $u(0) = c_1$ and there is no terminal condition, and then the cases where there are conditions at both boundaries, such as $u(0) = c_1, u(T) = c_2$.

Following this, we shall consider the more general problem of minimizing the functional

$$J(u, v) = \int_0^T (u^2 + v^2)\, dt, \qquad (4.1.2)$$

where the functions u and v are connected by the linear differential equation

$$u' = au + v, \qquad u(0) = c. \qquad (4.1.3)$$

We shall take T to be finite unless specifically stated otherwise, and all functions and parameters appearing are taken to be real. The possibility of using the method of successive approximations to handle the more important problem of minimizing

$$K(u, v) = \int_0^T g(u, v) \, dt, \qquad (4.1.4)$$

where u and v are related by the nonlinear equation

$$u' = h(u, v), \qquad u(0) = c, \qquad (4.1.5)$$

leads us then to consider the minimization of an expression such as

$$K(u) = \int_0^T (u'^2 + g(t)u^2) \, dt, \qquad (4.1.6)$$

under various hypotheses concerning the function $g(t)$. This investigation is intimately related to Sturm-Liouville theory as we will briefly indicate without allowing ourselves to be diverted from our principal goals.

4.2. Does a Minimum Exist?

The first question we must face, generally one of the thorniest in the calculus of variations, is that of determining the class of functions that we allow as candidates for the minimization of the functional

$$J(u) = \int_0^T (u'^2 + u^2) \, dt. \qquad (4.2.1)$$

Do we wish to consider all possible functions $u(t)$ defined over $[0, T]$, satisfying the condition $u(0) = c$? Obviously the answer is negative. To begin with, we can only admit functions for which $J(u)$ is defined. This does not end the matter, since we must now state what type of integral we are using. Is it a Riemann integral, a Lebesgue integral, or one of more exotic variety?

It turns out that the Riemann integral is satisfactory for our present purposes since the minimizing function, as we will show, is a very well-behaved function of t. However, for our subsequent purposes in Chapter 9, where we want to initiate the reader into the elements of the application of functional analysis to control theory, it is just as well to start by supposing that the integrals involved are Lebesgue integrals. We use only the elementary properties at this stage. Consequently, the reader unversed in real variable theory can without loss consider all integrals to be Riemann integrals.

Hence, our fundamental restriction is that $u(t)$ is a function with the property that it possesses a derivative $u'(t)$ whose square is Lebesgue-integrable over $[0, T]$. This fact is occasionally written: $u' \in L^2[0, T]$. A consequence of this is that $u(t)$, as the integral of $u'(t)$, is a continuous function with a derivative almost everywhere in $[0, T]$. In particular, $u \in L^2[0, T]$.

The function $u'(t)$ need not be continuous. The question of the admissible class of control functions is important both from the mathematical point of view and from that of the control engineer. The mathematician wants a class of functions large enough to ensure that it contains a minimizing function, while the engineer is interested in a class of functions corresponding to feasible and reasonable control actions. In a small number of cases, the two classes coincide.

In order to keep the presentation on the desired elementary path, we are going to follow a very circumscribed route. Rather than tackle the problem of the existence of a minimizing function directly, we will first derive a necessary condition that any such function must satisfy. This necessary condition is the celebrated differential equation of Euler.

It is not difficult, as we show, to establish the fact that this equation possesses a unique solution. Next we demonstrate, by means of a simple direct calculation, that the function obtained in this fashion yields the absolute minimum of $J(u)$ for all admissible u.

This circuitous approach enables us to provide a completely rigorous discussion of the variational problem with a minimum of analytic background. Furthermore, it provides a rigorous foundation for the dynamic programming approach of Chapter 5. Finally, all the steps in the argument have been carefully constructed so as to generalize to the multidimensional case, granted a modicum of vector-matrix manipulation. The essentials of a vector-matrix treatment of systems of linear differential equations will be given in Chapter 6, in preparation for the contents of Chapters 7 and 8.

EXERCISES

1. Show that if u is admissible, then $c_1 u$ is admissible for any scalar a_1.
2. Show that if u is admissible, then u^2 is admissible.
3. Show that $u + v$ is admissible if u and v are separately admissible. (*Hint:* $(u + v)^2 \leq 2(u^2 + v^2)$.)
4. Show that $\int_0^t u \, dt_1$ and $\int_t^T u \, dt_1$ are admissible if u is admissible.
5. Show that $u + \int_0^t \left[\int_t^T u \, ds \right] dt_1$ is admissible if u is admissible.
6. Using the fact that $\int_0^T (c_1 f + c_2 g)^2 \, dt$, where c_1 and c_2 are constants, is a positive definite quadratic form if f and g are linearly independent, establish the Cauchy-Schwarz inequality

$$\left(\int_0^T fg \, dt \right)^2 \leq \left(\int_0^T f^2 \, dt \right) \left(\int_0^T g^2 \, dt \right).$$

7. Show that

$$\left(\int_0^t u \, dt_1 \right)^2 \leq t \int_0^t u^2 \, dt_1.$$

8. Establish the Minkowski inequality

$$\left(\int_0^T (f + g)^2 \, dt \right) \leq \left(\int_0^T f^2 \, dt \right)^{1/2} + \left(\int_0^T g^2 \, dt \right)^{1/2}.$$

4.3. The Euler Equation

Let w denote a hypothetical solution to the minimization of the functional $J(u)$, as given in (4.1.1), and let z denote another function of t with the property that $J(z)$ exists. Take ε to be a scalar parameter and consider the expression

$$J(w + \varepsilon z) = \int_0^T [(w' + \varepsilon z')^2 + (w + \varepsilon z)^2] \, dt \tag{4.3.1}$$

as a function of the scalar variable ε. If w is indeed the function that yields the absolute minimum of $J(u)$, then $J(w + \varepsilon z)$ as a function of the scalar parameter ε must have an absolute minimum at $\varepsilon = 0$. Hence, using the usual result of calculus, a necessary condition is

$$\frac{d}{d\varepsilon} J(w + \varepsilon z) \bigg|_{\varepsilon = 0} = 0. \tag{4.3.2}$$

This actually does not require calculus since $J(w + \varepsilon z)$, as we see from (4.3.1), is a quadratic in ε,

$$J(w + \varepsilon z) = \int_0^T (w'^2 + w^2)\, dt + 2\varepsilon \int_0^T (w'z' + wz)\, dt$$

$$+ \varepsilon^2 \int_0^T (z'^2 + z^2)\, dt. \tag{4.3.3}$$

We see then that the variational condition derived from (4.3.2) is

$$\int_0^T (w'z' + wz)\, dt = 0, \tag{4.3.4}$$

for all z such that $J(z)$ exists.

If we knew that w had a second derivative, we would proceed in the following fashion. Integrating by parts, we obtain from (4.3.4) the relation

$$w'z \Big|_0^T + \int_0^T z(-w'' + w)\, dt = 0. \tag{4.3.5}$$

This is not a legitimate step since we do not know that w'' exists. Let us, however, forge ahead boldly. Since the left-hand side must be zero for all admissible z, we suspect that

$$-w'' + w = 0, \tag{4.3.6}$$

and that $w'(T) = 0$. Since $w + \varepsilon z$ as an admissible function satisfies the condition $w(0) + \varepsilon z(0) = c$, we see that $z(0) = 0$. Hence, we obtain no condition on $w'(0)$.

Equation (4.3.6) is the Euler equation associated with this variational problem.

Observe that the desired solution satisfies a two-point boundary condition

$$w(0) = c, \qquad w'(T) = 0. \tag{4.3.7}$$

Consequently, it is not immediately obvious that there exists a function $w(t)$ with the required properties. Nor at this point is it clear that this function serves any useful purpose in the sense of furnishing the desired minimum value.

4.4. A Fallacious Argument

In the foregoing section, following (4.3.5), we said, "... we sus-pect...." Why not proceed as follows? If

$$\int_0^T z(-w'' + w)\, dt = 0 \tag{4.4.1}$$

for *all* admissible z, set $z = -w'' + w$. Then (4.4.1) yields

$$\int_0^T (-w'' + w)^2\, dt = 0, \tag{4.4.2}$$

whence $-w'' + w = 0$, the desired result.

The fallacy resides in the fact that $z = -w'' + w$ need not be an admissible function. At the moment, all we know about w' is that $w' \in L^2[0, T]$, and, as pointed out above, we are not even sure that w'' exists.

Nevertheless, the conclusion, a special case of the fundamental lemma of the calculus of variations, is valid. For our present objective, that of minimizing $J(u)$, all we need is the Euler equation, not a rigorous derivation. Hence, we do not provide any rigorous validation of the step from (4.3.5) to (4.3.6). As we see below, there is a simple way of making an argument of this nature rigorous.

4.5. Haar's Device

Returning (4.3.4), let us regard the derivative w' as the fundamental function and integrate the term $\int_0^T wz\, dt$ by parts as follows:

$$\int_0^T wz\, dt = \left(-z \int_t^T w\, ds\right)\Big|_0^T + \int_0^T z'\left(\int_t^T w\, ds\right) dt. \tag{4.5.1}$$

This is an ingenious device from the standpoint of the calculus of variations, but a very natural step as far as control theory is concerned, since w' corresponds to the control law. The integrated term vanishes at

$t = 0$ because $z(0) = 0$, and obviously at $t = T$. Hence, we have

$$\int_0^T z' \left(w' + \int_t^T w \, ds \right) dt = 0 \qquad (4.5.2)$$

for all z for which $J(z)$ exists. But the function $w' + \int_t^T w \, ds$ is now an admissible choice for z'. Hence, the simple technique described in Section 4.4 is rigorously applicable. If (4.5.2) holds for all $z' \, \varepsilon \, L^2[0, T]$, then we can set $z' = w' + \int_t^T w \, ds$ and deduce that

$$w' + \int_t^T w \, ds = 0 \qquad (4.5.3)$$

for $0 \le t \le T$. Actually, from

$$\int_0^T \left[w' + \int_t^T w \, ds \right]^2 dt = 0,$$

we only deduce that (4.5.3) holds almost everywhere. As the next argument shows, this is no essential restriction. From this it follows that w', as the integral of a continuous function, is differentiable, and thus, differentiating, that

$$w'' - w = 0, \qquad (4.5.4)$$

the Euler equation once again.

Furthermore, (4.5.3) supplies the missing boundary condition, to wit,

$$w'(T) = 0. \qquad (4.5.5)$$

Let it be noted, however, that we have not as yet established the existence of a minimizing function $w(t)$.

4.6. Solution of the Euler Equation

Consider now the equation

$$u'' - u = 0, \qquad u(0) = c, \qquad u'(T) = 0. \qquad (4.6.1)$$

The general solution of the differential equation is

$$u = c_1 e^t + c_2 e^{-t}. \qquad (4.6.2)$$

Using the boundary conditions, we have the two equations

$$c = c_1 + c_2,$$
$$0 = c_1 e^T - c_2 e^{-T}$$

(4.6.3)

to determine the coefficients c_1 and c_2. Solving, we obtain the expression

$$u = c\left(\frac{e^{t-T} + e^{-(t-T)}}{e^{-T} + e^T}\right) = c\,\frac{\cosh(t-T)}{\cosh T}.$$

(4.6.4)

This is the unique solution of (4.6.1). It exists for all $T > 0$ since the denominator $e^{-T} + e^T$ is never zero. We have thus shown that there is one and only one function satisfying the necessary condition expressed by the Euler equation subject to the two-point boundary condition.

4.7. Minimizing Property of the Solution

Let us now show that the function obtained in this fashion yields the minimum value of J. Consider

$$J(u + w) = \int_0^T [(u' + w')^2 + (u + w)^2]\,dt$$

$$= \int_0^T (u'^2 + u^2)\,dt + 2\int_0^T (u'w' + uw)\,dt + \int_0^T (w'^2 + w^2)\,dt$$

(4.7.1)

for any function w such that $w(0) = 0$ (to ensure that $u(0) + w(0) = c$), and such that $w' \in L^2[0, T]$. The middle term on the right disappears upon integration by parts,

$$\int_0^T (u'w' + uw)\,dt = [wu']_0^T + \int_0^T w(-u'' + u)\,dt = 0.$$

(4.7.2)

Hence

$$J(u + w) = J(u) + J(w) > J(u)$$

(4.7.3)

unless w is identically zero.

Thus we have established by a direct calculation that J has a minimum value assumed by the unique solution of the Euler equation determined by (4.6.4). The fact that the Euler equation was obtained by a formal procedure is no longer of any consequence.

4.8. Alternative Approach

In Section 4.6 we used the fact that the Euler equation (4.6.1) could be solved explicitly. This enabled us to verify directly that a solution existed for all $T > 0$ and that it was unique. Let us now employ a more powerful method which is applicable to the more common cases where we do not possess an explicit analytic representation for the solution.

Let v and w be two solutions of the Euler equation, (4.6.1). Then $u = v - w$ satisfies the differential equation and the conditions $u(0) = 0$, $u'(T) = 0$.

Suppose that the equation

$$u'' - u = 0 \qquad (4.8.1)$$

possessed a nontrivial solution satisfying the boundary conditions $u(0) = 0$, $u'(T) = 0$. Then, multiplying by u and integrating over $[0, T]$,

$$\int_0^T u(u'' - u)\, dt = 0. \qquad (4.8.2)$$

Integrating by parts, this yields

$$uu']_0^T - \int_0^T (u'^2 + u^2)\, dt = 0. \qquad (4.8.3)$$

The integrated terms vanish, and we are left with the relation

$$\int_0^T (u'^2 + u^2)\, dt = 0. \qquad (4.8.4)$$

This can hold only if u and u' are identically zero.

4.9. Asymptotic Control

In the calculus of variations, emphasis is customarily placed upon establishing the existence of a solution over some T-interval, however small. In control theory, it is natural to wish to examine the nature of the solution for all $T > 0$. In particular, we are interested in the behavior of the control law as $T \to \infty$. Just as in the study of descriptive processes, we can expect a great deal of simplification as a steady-state regime takes over. As we shall see, this is indeed the case.

Let us examine the explicit form of the solution given in (4.6.4),

$$
u = c \frac{\cosh (t - T)}{\cosh T} = c \frac{[e^{t-T} + e^{-(t-T)}]}{[e^T + e^{-T}]}
$$

$$
= \frac{ce^{-t}}{(1 + e^{-2T})} + \frac{ce^{t-2T}}{(1 + e^{-2T})}. \tag{4.9.1}
$$

If $0 \leq t \leq T$, the second term on the right is uniformly bounded by the quantity $|c|e^{-T}$. Hence,

$$
\left| u - \frac{ce^{-t}}{(1 + e^{-2T})} \right| \leq |c|e^{-T}, \tag{4.9.2}
$$

and thus

$$
|u - ce^{-t}| \leq 2|c|e^{-T}
$$

for $0 \leq t \leq T$. From the expression

$$
u' = \frac{c \sinh (t - T)}{\cosh T}, \tag{4.9.3}
$$

we obtain the same way

$$
\left| u' + \frac{ce^{-t}}{(1 + e^{-2T})} \right| \leq |c|e^{-T},
$$

$$
|u' + ce^{-t}| \leq 2|c|e^{-T}. \tag{4.9.4}
$$

Thus, if T is large, we have the approximate relation

$$
u' \cong -u \tag{4.9.5}
$$

uniformly valid in $0 \leq t \leq T$. This is exactly the type of simple control law we were seeking.

We have thus succeeded in converting a control process that was time-dependent, in the sense that we considered u' as a function of t, to a control process that is state-dependent, u' depending on u. We may not always want to do this. What we do want is the flexibility to obtain whatever analytic representation of optimal control that is most convenient for the subsequent mathematical or engineering applications.

4.10. Infinite Control Process

Suppose we wish to consider a control process over an unbounded time interval. Take, for example, the problem of minimizing the functional

$$J_1(u) = \int_0^\infty (u'^2 + u^2)\, dt, \qquad (4.10.1)$$

where again $u(0) = c$.

The first point to mention is that the class of admissible functions has changed. We must impose the additional restrictions that $\int_0^\infty u'^2\, dt$ and $\int_0^\infty u^2\, dt$ converge.

The second point to clarify is that of determining whether a solution to the foregoing minimization problem can be obtained as a limit of the solution to the finite control process as $T \to \infty$. The latter problem is by far the more important of the two questions.

We shall leave it to the reader to examine these questions in the form of exercises.

EXERCISES

1. Show formally that the Euler equation is $u'' - u = 0$ subject to the condition that $u(0) = c$ and that u vanishes at infinity. Does this equation have a unique solution and does this solution furnish the absolute minimum of $J_1(u)$?

2. Show that the solution to this two-point boundary problem may be obtained as the limit of the solution to the finite process as $T \to \infty$. Is the limit uniform in any finite t-interval?

3. What is the limiting form of the control law?

4. If the steady-state control law $u' = -u$ is used in the finite process over $[0, T]$, how much of an error results?

5. Prove without any analytic calculation that $\lim_{T \to \infty} \min_u J(u)$ exists by means of the following arguments: (1) As a function of T, $\min_u J(u)$ is monotone increasing. (2) Using the admissible function $u = ce^{-t}$, we see that

$$\min_u J(u) \le \int_0^T [c^2 e^{-2t} + c^2 e^{-2t}]\, dt < 2c^2 \int_0^\infty e^{-2t}\, dt = c^2.$$

Hence, $\min_u J(u)$ is uniformly bounded for $T \ge 0$.

6. What are power series expansions for the minimizing function u and $\min_u J(u)$ for small T? Do these power series expansions converge for all $T \geq 0$? If not, why not?

7. Use the trial function (a term equivalent to "admissible solution") $u = c + b_1 t$ and calculate

$$J(u) = \int_0^T [b_1{}^2 + (c + b_1 t)^2] \, dt.$$

Minimizing over b_1, do we obtain the same coefficient b_1 as in the power series expansion of the minimizing function?

8. Can we determine b_1 immediately by the condition $u'(T) = 0$?

9. If we take $u = c + b_1 t + \cdots + b_M t^M$, do we obtain a more accurate answer by minimizing $J(u) = J(b_1, b_2, \ldots, b_M)$ over the variables $b_1, b_2, \ldots, b_M$ or by minimizing over $b_1, b_2, \ldots, b_{M-1}$ with b_M determined by the condition $u'(T) = 0$?

10. Consider $\min J(b_1, b_2, \ldots, b_M)$ as a function of M for $M = 1, 2, \ldots$. Calculate the values for $M = 1, 2, 3$ for $T = 1$. Use an extrapolation formula to estimate the limiting values as $M \to \infty$. How much of an error is there? (See Section 8.13.)

11. Calculate an approximate value for the minimizing function $u(t)$ in this way.

4.11. The Minimum Value of $J(u)$

In order to calculate the minimum value of $J(u)$, a quantity of some significance to us subsequently, we proceed as follows. From the equation

$$u'' - u = 0, \qquad u(0) = c, \qquad u'(T) = 0, \tag{4.11.1}$$

we have

$$\int_0^T u(u'' - u) \, dt = 0 \tag{4.11.2}$$

and thus, integrating by parts,

$$[uu']_0^T - \int_0^T (u'^2 + u^2) \, dt = 0. \tag{4.11.3}$$

Hence, for the minimizing function u,

$$J(u) = \int_0^T (u'^2 + u^2)\, dt = [uu']_0^T = -u(0)u'(0) = -cu'(0). \qquad (4.11.4)$$

Rather than use the explicit form of the minimizing function, let us first use the expression for $u(t)$ in terms of principal solutions, namely

$$u(t) = c\left[u_1(t) - \frac{u_2(t)}{u_2'(T)} u_1'(T) \right]. \qquad (4.11.5)$$

We have from this

$$u'(0) = -\frac{u_1'(T)}{u_2'(T)}\, c. \qquad (4.11.6)$$

Hence

$$\min_u J(u) = c^2 \frac{u_1'(T)}{u_2'(T)}. \qquad (4.11.7)$$

This proves incidentally that $u_1'(T)/u_2'(T)$ is positive. Furthermore, since $u_2'(0) = 1$ and $u_2'(T)$ is never zero, we deduce that $u_2'(T) > 0$ and thus that $u_1'(T) > 0$.

Returning to the explicit expressions for u_1 and u_2,

$$u_1 = \cosh t, \qquad u_2 = \sinh t, \qquad (4.11.8)$$

we see that

$$\min_u J(u) = c^2 \frac{\sinh T}{\cosh T} = c^2 \tanh T. \qquad (4.11.9)$$

Since

$$\tanh T = \frac{e^T - e^{-T}}{e^T + e^{-T}} = \frac{1 - e^{-2T}}{1 + e^{-2T}}, \qquad (4.11.10)$$

we see that

$$\lim_{T \to \infty} \min_u J(u) = c^2. \qquad (4.11.11)$$

EXERCISES

1. Is it true that $\lim_{T \to \infty} (\min_u J(u)) = \min_u (\lim_{T \to \infty} J(u))$?

2. Consider the problem of minimizing

$$J_\lambda(u) = \int_0^T (u'^2 + \lambda u^2) \, dt,$$

where $\lambda \geq 0$ and $u(0) = c$. Write $\min_u J_\lambda(u) = f(\lambda)$. Is $f(\lambda)$ continuous for $\lambda \geq 0$?

3. Make the change of variable $t = as$ for an appropriate choice of a and show that we can obtain the minimum of $J_\lambda(u)$ in terms of the minimum of $J(u)$.

4. Consider the functional

$$K_\lambda(u) = \int_0^T (\lambda u'^2 + u^2) \, dt$$

where $\lambda \geq 0$. Let $g(\lambda) = \min_u K_\lambda(u)$. Is $g(\lambda)$ continuous for $\lambda \geq 0$?

5. What are the limiting forms of the minima of $J_\lambda(u)$ and $K_\lambda(u)$ as $\lambda \to \infty$? What are the limiting forms of the minimizing functions?

4.12. Two-Point Constraints

In many cases of interest, the minimizing function is constrained at both boundaries,

$$u(0) = c_1, \qquad u(T) = c_2. \tag{4.12.1}$$

Proceeding as before, we see that a plausible necessary condition for $u(t)$ to minimize $J(u)$ is that it satisfy the Euler equation

$$u'' - u = 0, \tag{4.12.2}$$

subject to (4.12.1).

The unique solution to (4.12.2) and (4.12.1) is readily obtained in explicit analytic form, and the argument given above suffices to show that this function yields the absolute minimum of $J(u)$ within the admissible class of functions.

EXERCISES

1. Show that $\min_u J(u) = c_1{}^2 r_1(T) + 2c_1 c_2 r_2(T) + c_2{}^2 r_3(T)$, where r_1, r_2, r_3 as indicated are functions of T alone. What relation exists between r_1 and r_3?

2. Show that r_1, r_2, r_3 approach finite limits as $T \to \infty$. What are these limits?

3. Show that r_1 and r_3 are monotone increasing as $T \to \infty$. Does this hold for r_2?

4. What is the behavior of r_1, r_2, r_3 as $T \to 0$?

5. As $T \to 0$, show that $u \cong c_1 + (t(c_2 - c_1))/T$, $u' \cong (c_2 - c_1)/T$.

6. What is the minimum value of $J(u)$ taken over all quadratic functions $u = c_1 + b_1 t + b_2 t^2$ passing through (T, c_2)?

7. Consider the condition $a_1 u(T) + a_2 u'(T) = 0$. Obtain the explicit analytic representation of the extremal function $u(t)$ and determine its limiting behavior as $a_1 \to \infty$ and then as $a_2 \to \infty$.

8. Consider the problem of minimizing

$$J(u, \varepsilon) = \int_0^T [\varepsilon u'^2 + u^2]\, dt,$$

where $\varepsilon > 0$ and $u(0) = c_1$, $u(T) = c_2$. What is the behavior of $\min_u J(u, \varepsilon)$ and the extremal function u as $\varepsilon \to 0$?

9. Does $\lim_{\varepsilon \to 0} \min_u J(u, \varepsilon) = \min_u \lim_{\varepsilon \to 0} J(u, \varepsilon)$?

10. Does $\lim_{T \to \infty} \lim_{\varepsilon \to 0} \min_u J(u, \varepsilon) = \lim_{\varepsilon \to 0} \lim_{T \to \infty} \min_u J(u, \varepsilon)$?

4.13. Terminal Control

In a number of important control processes, the criterion for efficiency is a mixture of the history of the process and the final state of the system. As an example of this, consider the problem of minimizing the functional

$$J_\lambda(u) = \int_0^T [u'^2 + u^2]\, dt + \lambda u(T)^2, \qquad (4.13.1)$$

over all admissible functions for which $u(0) = c$. Here $\lambda \geq 0$.

We derive a necessary condition, as before. Let $\bar{u}(t)$ be a minimizing function and write $u = \bar{u} + \varepsilon v$. Setting the derivative with respect to ε equal to zero, we obtain the relation

$$\int_0^T [\bar{u}'v' + \bar{u}v] \, dt + \lambda \bar{u}(T)v(T) = 0. \qquad (4.13.2)$$

Integrating by parts, we have

$$\int_0^T v[\bar{u}'' - \bar{u}] \, dt - v(T)[\bar{u}'(T) + \lambda \bar{u}(T)] = 0. \qquad (4.13.3)$$

Hence, we suspect that the appropriate conditions are

$$u'' - u = 0, \qquad u(0) = c, \qquad u'(T) + \lambda u(T) = 0. \qquad (4.13.4)$$

We leave it to the reader to verify that (4.13.4) has a unique solution that provides the absolute minimum of $J_\lambda(u)$.

EXERCISES

1. Compute $\min_u J_\lambda(u)$ and determine its analytic structure as a function of λ.
2. Determine $\lim_{\lambda \to \infty} J_\lambda(u)$. Is it equal to the minimum of $J(u)$ subject to the condition $u(T) = 0$?
3. Is it true that $\min_u J_\lambda(u)$ is determined by a knowledge of the minimum for three distinct values of λ?
4. Determine $\lim_{T \to \infty} \min_u J_\lambda(u)$. Is $\lim_{T \to \infty} \lim_{\lambda \to \infty} \min_u J_\lambda(u) = \lim_{\lambda \to \infty} \lim_{T \to \infty} \min_u J_\lambda(u)$?
5. Consider the problem of minimizing

$$K(u) = \int_0^T (u'^2 + u^2) \, dt + \lambda u'(T)^2$$

over all admissible functions satisfying $u(0) = c$. Show that $\min_u K(u) = \min_u J(u)$. (*Hint:* Recall the definition of an admissible function!)
6. Consider the minimum of

$$J_a(u) = \int_0^T (u'^2 + auu' + u^2) \, dt$$

over all u such that $u(0) = c$, where a is a constant. What is the behavior of $\min_u J_a(u)$ as a becomes more and more negative? Is there a critical value of a, and if so, what is it?

7. What is the minimum of $\int_0^T (u'^2 + u^2)\, dt$ over all u subject to $u(0) = c_1, u(t_0) = c_2$, where $0 < t_0 \le T$?
8. What is the minimum of $\int_0^T (u'^2 + u^2)\, dt + \lambda u(t_0)^2$, where $0 < t_0 \le T$ and $\lambda \ge 0$?
9. What is the minimum of $\int_0^T (u'^2 + u^2)\, dt$, where $u(0) = c_1, u(T) + \lambda u'(T) = c_2$? Is the minimum a continuous function of λ for $0 \le \lambda \le \infty$? What is the limiting behavior as $\lambda \to \infty$?

4.14. The Courant Parameter

Frequently, in the calculus of variations and in applying dynamic programming, it is inconvenient to invoke the terminal condition $u(T) = c_2$. Instead, we think of paying a penalty for the deviation of $u(T)$ from c_2. Consider then the functional

$$J_\lambda(u) = \int_0^T [u'^2 + u^2]\, dt + \lambda(u(T) - c_2)^2, \qquad (4.14.1)$$

where λ is a positive parameter called the Courant parameter.

From what has preceded, it is plausible that the solution of the problem of minimizing $J_\lambda(u)$ subject to $u(0) = c_1$ approaches the extremal function associated with $J(u)$ as $\lambda \to \infty$. We leave this for the reader to demonstrate.

EXERCISE

1. Consider the problem of minimizing

$$K_\lambda(u) = \int_0^T (u'^2 + u^2)\, dt + \lambda\left(\int_0^T u\, dt - a\right)^2$$

over all admissible $u(t)$ subject to $u(0) = c$, where $\lambda \ge 0$. What is the limiting behavior of the minimum value as $\lambda \to \infty$?

4.15. Successive Approximations

As we indicated in Chapter 1, control processes can be quite complex. Consequently, in order to test our concepts and mathematical tools against the intricacies of control, we begin with quite simple mathematical models of control processes. This motivates our consideration initially of the problem of minimizing the function

$$J(u) = \int_0^T g(u, u') \, dt, \qquad u(0) = c. \tag{4.15.1}$$

In order to display some of the techniques available for the treatment of this problem and to analyze some of the results in detail, we considered first the case where g was a quadratic function of u and u'.

These results are clearly of interest in themselves. They are, moreover, of major importance in enabling us to tackle the more general problem posed above. In studying general control policies, we have a number of objectives. To begin with, we always want to determine optimal control, if possible. In a number of situations, however, the problem is not so much this as it is that of changing current operational procedures step-by-step to improve the performance of the system. Another way of putting this is that we are very much interested in methods of successive approximations that steadily upgrade the behavior. In connection with the control of an operating system, a realistic appraisal of the time and effort required to change old procedures and institute new ones forces us to consider this more general type of control process.

Leaving aside this important class of problems, let us consider a straightforward application of the method of successive approximations to the problem of minimizing the function in (4.15.1). Let $u_0(t)$ be an initial approximation to the solution to the problem of the minimization of (4.15.1). To obtain an improved estimate, we expand $g(u, u')$ around u_0, retaining the zeroth-, first-, and second-order terms,

$$g(u, u') = g(u_0, u_0') + (u - u_0)g_{u_0} + (u' - u_0')g_{u_0'}$$

$$+ g_{u_0 u_0} \frac{(u - u_0)^2}{2} + g_{u_0 u_0'}(u - u_0)(u' - u_0')$$

$$+ g_{u_0' u_0'} \frac{(u' - u_0')^2}{2} + \text{higher-order terms}$$

$$= g_2(u, u'; u_0, u_0') + \text{higher-order terms}, \tag{4.15.2}$$

where g_2 denotes the sum of terms of second order or less.

We next consider the approximate variational problem

$$\min_u \int_0^T g_2(u, u'; u_0, u_0')\, dt, \qquad u(0) = c_1. \tag{4.15.3}$$

This determines a new function u_1. We form $g_2(u, u'; u_1, u_1')$ and continue in this fashion. In Volume II, we will justify this method. At the moment, we merely wish to indicate one way in which quadratic variational problems with time-dependent integrands occur. The interesting point is that a problem of time-independent type has led us to a problem involving time-dependence by way of the appearance of the previous approximation, u_0, u_0'.

4.16. min $\int_0^T [u'^2 + g(t)u^2]\, dt$

In order to illustrate the methods that can be employed to handle the time-dependent case, which are precisely the same as those for the time-independent case, we shall consider the problem of minimizing

$$K(u) = \int_0^T [u'^2 + g(t)u^2]\, dt, \tag{4.16.1}$$

where $u(0) = c$. Let us begin once again by proceeding formally. The Euler equation, obtained as before, is

$$u'' - g(t)u = 0, \tag{4.16.2}$$

subject to the two conditions $u(0) = c$, $u'(T) = 0$.

If $g(t)$ is a function of general form, (4.16.2) cannot be solved explicitly in terms of the elementary functions. Consequently, there is no easy way of determining when a solution exists. What we do know is that a solution does not always exist.

Consider, for example, the case where $g(t) = -1$ and $T = \pi/2$. The Euler equation is now

$$u'' + u = 0, \tag{4.16.3}$$

whose general solution is

$$u = c_1 \cos t + c_2 \sin t. \tag{4.16.4}$$

The condition at $t = 0$ yields $c_1 = c$. The condition at $t = \pi/2$ yields the equation

$$0 = -c \sin \frac{\pi}{2} + c_2 \cos \frac{\pi}{2}. \qquad (4.16.5)$$

Since $\cos \pi/2 = 0$, c_2 cannot be determined. Furthermore, the equation cannot possibly hold if $c \neq 0$.

Nevertheless, the minimization problem makes sense. Indeed, as we shall indicate below, using Sturm-Liouville theory, we can establish the fact that

$$\min_{u} \int_0^{\pi/2} (u'^2 - u^2) \, dt = 0, \qquad (4.16.6)$$

and the minimum is assumed for $u(t) = c \cos t$.

On the other hand, if we consider the case where $g(t) = -4$, $T = \pi/2$, we see that the Euler equation

$$u'' + 4u = 0, \qquad u(0) = c, \qquad u'\left(\frac{\pi}{2}\right) = 0 \qquad (4.16.7)$$

has a unique solution

$$u = c \cos 2t. \qquad (4.16.8)$$

Yet the functional $\int_0^{\pi/2} (u'^2 - 4u^2) \, dt$ possesses no minimum value if c is unrestricted. If $c = 0$, the unique solution of the Euler equation is $u = 0$. The functional can assume arbitrarily large negative values as we see upon setting $u = b \sin t$. We have

$$\int_0^{\pi/2} (b^2 \cos^2 t - 4b^2 \sin^2 t) \, dt = -3b^2 \int_0^{\pi/2} \sin^2 t \, dt, \qquad (4.16.9)$$

since

$$\int_0^{\pi/2} \sin^2 t \, dt = \int_0^{\pi/2} \cos^2 t \, dt.$$

Taking b arbitrarily large, we see that $\int_0^{\pi/2} (u'^2 - 4u^2) \, dt$ can assume arbitrarily large negative values.

4.17. Discussion

We trust that these simple examples demonstrate the fact that general variational problems can possess considerable features of complexity. In particular, one has to be quite careful about using the character of the solution of the Euler equation as a guide to the nature of the solution of the original minimization problem, or even as a guide to the existence of a minimizing function. Fortunately, as we will show, the intrinsic nature of control processes makes it easy for us to obtain definitive results.

4.18. The Simplicity of Control Processes

What considerably simplifies our task in the study of control processes is that "costs" are automatically positive. We have already used this property in an alternative approach to the Euler equation with constant coefficients. Let us now use the same method to establish

Theorem. *Assume that $g(t)$ is continuous for $t \geq 0$ and that*

$$J(u) = \int_0^T [u'^2 + g(t)u^2] \, dt \qquad (4.18.1)$$

is positive for any $T > 0$ if u is not identically zero. Then $J(u)$ possesses an absolute minimum over all functions $u' \in L^2[0, T]$ that satisfy the initial condition $u(0) = c$. The extremal function is determined as the unique solution of the Euler equation

$$u'' - g(t)u = 0, \qquad u(0) = c, \qquad u'(T) = 0. \qquad (4.18.2)$$

Proof. Let us begin by showing that (4.18.2) possesses a unique solution. Let u_1, u_2 be, respectively, the solutions of $u'' - g(t)u = 0$ determined by the initial conditions $u_1(0) = 1$, $u_1'(0) = 0$, $u_2(0) = 0$, $u_2'(0) = 1$, the principal solutions. Writing $u = c_1 u_1 + c_2 u_2$, and taking account of the boundary conditions, we see that $c_1 = c$ and that c_2 is determined by the equation

$$0 = cu_1'(T) + c_2 u_2'(T). \qquad (4.18.3)$$

To establish both existence and uniqueness, we must show that $u_2'(T) \neq 0$. The proof is by contradiction. Suppose $u_2'(T) = 0$. Then we possess a nontrivial solution of

$$u'' - g(t)u = 0, \qquad u(0) = 0, \qquad u'(T) = 0. \qquad (4.18.4)$$

Multiplying by u and integrating, we have

$$\int_0^T u(u'' - g(t)u)\, dt = 0. \qquad (4.18.5)$$

Integrating by parts, we have

$$uu']_0^T - \int_0^T (u'^2 + g(t)u^2)\, dt = 0. \qquad (4.18.6)$$

The boundary conditions cause the integrated terms to vanish. What remains in (4.18.6) contradicts our assumption concerning $J(u)$. Hence, $u_2'(T) \neq 0$, which means we can determine c_2 from (4.18.3). Since $u_2'(0) = 1$, we see that $u_2'(T) > 0$ for all $T > 0$.

Let $u(t)$ be the solution of (4.18.2) obtained in this fashion,

$$u = cu_1(t) - c\, \frac{u_2(t)u_1'(T)}{u_2'(T)}. \qquad (4.18.7)$$

As before, we can show that $J(v) > J(u)$ for any v distinct from u for which $v(0) = c$. Write $v = u + (v - u) = u + w$, where $w(0) = 0$. Then

$$J(u + w) = \int_0^T [(u' + w')^2 + g(t)(u + w)^2]\, dt$$

$$= J(u) + J(w) + 2\int_0^T [u'w' + g(t)uw]\, dt. \qquad (4.18.8)$$

Integrating by parts, we have

$$\int_0^T [u'w' + g(t)uw]\, dt = [u'w]_0^T + \int_0^T w[-u'' + g(t)u]\, dt = 0. \qquad (4.18.9)$$

Hence

$$J(u + w) = J(u) + J(w) > J(u) \qquad (4.18.10)$$

unless $w \equiv 0$. This completes the proof of the Theorem.

1. Show that $u(t) > 0$ for $T \geq t \geq 0$ if $c > 0$.

2. Consider the problem of minimizing

$$J_\lambda(u) = \int_0^T (u'^2 + u^2) \, dt + \lambda u^2(t_0),$$

where $0 < t_0 \leq T$, with $\lambda \geq 0$, and $u(0) = c$.

3. Consider the problem of minimizing

$$J(u) = \int_0^T [p(t)u'^2 + q(t)u^2] \, dt, \qquad u(0) = c.$$

4. Consider the problem of minimizing this functional subject to the conditions $u(0) = c_1$, $u(T) = c_2$.

4.19. Discussion

Examining the proof, we see that we have not required the full hypothesis that $J(u) > 0$ for all nontrivial u. What we have used is

(a) $J(u_2) > 0$,

(b) $J(w) > 0$ for all nontrivial w for which $w(0) = 0$.

(4.19.1)

Since (a) is a special case of (b), we see that (4.19.1b) is all that is required.

The simplest case in which (4.19.1b) holds is that where $g(t) \geq 0$ for $t \geq 0$. Let us also show that if T is small, (4.19.1b) is satisfied. We have

$$w(t) = \int_0^t w'(t_1) \, dt_1,$$

$$w^2(t) = \left(\int_0^t w'(t_1) \, dt_1 \right)^2 \leq t \int_0^t w'^2 \, dt_1,$$

(4.19.2)

upon using the Cauchy-Schwarz inequality. Hence,

$$|g(t)|w^2(t) \leq kt \int_0^t w'^2 \, dt_1,$$

(4.19.3)

where $k = \max_{0 \le t \le T} |g|$. Thus

$$\int_0^T |g| w^2 \, dt_1 \le k \int_0^T t \left(\int_0^t w'^2 \, dt_1 \right) dt$$

$$\le \frac{kT^2}{2} \int_0^T w'^2 \, dt_1$$

$$< \int_0^T w'^2 \, dt_1 \qquad (4.19.4)$$

if T is small. Hence $\int_0^T w'^2 \, dt_1 > \int_0^T |g| w^2 \, dt_1$ for T small, which means that $\int_0^T (w'^2 + g w^2) \, dt > 0$ for T small and w' not identically zero.

4.20. The Minimum Value of $J(u)$

From (4.18.4), we have

$$0 = \int_0^T u(u'' - g(t)u) \, dt = [uu']_0^T - \int_0^T (u'^2 + g(t)u^2) \, dt. \qquad (4.20.1)$$

Hence,

$$\min_u J(u) = -u(0)u'(0)$$

$$= -c\left[-c \frac{u_1'(T)}{u_2'(T)} \right] = c^2 \frac{u_1'(T)}{u_2'(T)}. \qquad (4.20.2)$$

The value of $u'(0)$ is obtained using (4.18.7)

EXERCISE

1. Show directly, without any calculation of the solution, that $\min_u J(u)$ must have the form $c^2 r(T)$, provided that the minimum exists.

4.21. A Smoothing Process

An interesting class of control processes involving time-dependent integrands arises in connection with "smoothing." Suppose that $h(t)$ is a function with some undesired oscillatory behavior and that we wish to approximate to it by a function $u(t)$ whose rate of change is smaller, on the average. One way to approach this problem is to determine the function u that minimizes

$$J(u) = \int_0^T [u'^2 + (u - h(t))^2] \, dt. \qquad (4.21.1)$$

Let us suppose that we impose an initial condition $u(0) = c$.

It is easy to verify that the Euler equation is

$$u'' - u = -h(t), \qquad u(0) = c, \qquad u'(T) = 0. \qquad (4.21.2)$$

The solution of this is

$$u = cu_1 - \int_0^T k(t, t_1) h(t_1) \, dt_1, \qquad (4.21.3)$$

where $k(t, t_1)$ is the appropriate Green's function and u_1 is the solution of the homogeneous equation satisfying $u_1(0) = 1$, $u_1'(T) = 0$.

EXERCISES

1. Show that the solution of (4.21.2) yields the absolute minimum of $J(u)$. What is this minimum value?
2. If the initial condition c is left free, what is the value which minimizes?
3. Obtain the solution for the case where $h(t) = 1$, $0 \le t \le b < T$, $h(t) = 0$, $b < t \le T$.
4. What is the function that minimizes

$$J(u, \lambda_1, \lambda_2) = \int_0^T [u'^2 + \lambda_1(u - h_1(t))^2 + \lambda_2(u - h_2(t))^2] \, dt,$$

where $\lambda_1, \lambda_2 \ge 0$, $u(0) = c$?
5. Show that the minimum value of

$$K(u) = \int_0^T [u'^2 + u^2 - 2h(t)u] \, dt$$

subject to $u(0) = u(T) = 0$ is the quadratic functional

$$Q(h) = -\int_0^T \int_0^T k(t, t_1) h(t) h(t_1) \, dt \, dt_1.$$

6. From this, show that $k(t, t_1)$ is a positive definite kernel, that is,

$$\int_0^T \int_0^T k(t, t_1) h(t) h(t_1) \, dt \, dt_1 > 0$$

for all nontrivial integrable $h(t)$. (*Hint:* $u = 0$ is an admissible function in the foregoing variational problem.)

4.22. Variation-Diminishing Property of Green's Function

In view of the derivation of the expression

$$v(t) = \int_0^T k(t, t_1) h(t_1) \, dt_1, \tag{4.22.1}$$

we would expect that $v(t)$ would be a smoother function than h, in the sense that it is nonpositive if $h(t)$ is nonnegative, and that it has no greater number of oscillations than $h(t)$. The nonpositivity can be demonstrated by direct calculation, using the explicit analytic form of $k(t, t_1)$.

It is, however, much more in the spirit of our preceding investigations to obtain the result without calculation. Furthermore, the method we present can be applied to more general equations. As we know, the Euler equation associated with the functional

$$J(u) = \int_0^T [u'^2 + u^2 + 2f(t)u] \, dt, \tag{4.22.2}$$

with the conditions $u(0) = 0$, $u'(T) = 0$, is

$$u'' - u = f(t), \qquad u(0) = u'(T) = 0. \tag{4.22.3}$$

The solution of this equation is $\int_0^T k(t, t_1) f(t_1) \, dt_1$. Let us now consider the case where $f(t) \geq 0$ in $[0, T]$.

Let us demonstrate directly from (4.22.2) that the function $\bar{u}$ uniquely determined by (4.22.1) that furnishes the absolute minimum of $J(u)$ must be nonpositive. The proof is by contradiction. Suppose that

$\bar{u} > 0$ in (a, b); see Fig. 4.1. Consider the new function determined by reflecting $\bar{u}$ in the t-axis in (a, b), but leaving it otherwise unchanged in the remainder of the interval; Fig. 4.2. The function $v(t)$ has a discontinuous derivative at $t = a$ and $t = b$, but this does not affect the existence

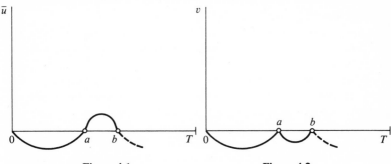

| Figure 4.1 | Figure 4.2 |

of $\int_0^T v'^2 \, dt$. The term $\int_0^T (v'^2 + v^2) \, dt$ has the same value as before. If, however, $f(t) > 0$ in (a, b), we have

$$\int_a^b f(t)v(t) \, dt < \int_a^b f(t)u(t) \, dt. \tag{4.22.4}$$

Hence

$$J(v) < J(\bar{u}), \tag{4.22.5}$$

a contradiction to the fact that $J(\bar{u})$ is the absolute minimum.

Thus the function

$$u(t) = \int_0^T k(t, t_1)f(t_1) \, dt_1 \tag{4.22.6}$$

is nonpositive for all $f(t) \geq 0$. This means that $k(t, t_1) \leq 0$ for $0 \leq t, t_1 \leq T$.

EXERCISE

1. Show in the same fashion that $u(t) = \int_0^T k(t, t_1)f(t_1) \, dt_1$ cannot have more variations in sign than $f(t)$. Hence, the Green's function $k(t, t_1)$ has the stated variation-diminishing property.

4.23. Constraints

We considered initially the problem of minimizing a functional representing the total cost of control, one part of which derived from the cost of deviation from desired performance and the other from the cost of exerting control.

In some cases, the situation is different. It is a matter of doing the best that one can, subject to a limitation on the resources available, the time available, the information available, and so on. These problems involving constraints are generally far more difficult than those we have so far encountered. There are, however, certain classes of variational problems subject to constraints that can be reduced in a simple and direct fashion to free variational problems. The general method is extremely powerful and quite significant. We shall introduce it in connection with a simple control process, as usual. Subsequently, in Chapter 9, we will present another version of the same method, which is much more powerful and flexible.

We wish to examine the problem of minimizing the functional

$$J(u) = \int_0^T u'^2 \, dt \qquad (4.23.1)$$

over all $u(t)$ for which $u(0) = c$, and for which

$$\int_0^T u^2 \, dt = k, \qquad (4.23.2)$$

where $k > 0$.

An intuitive way to approach this is the following. Let $\int_0^T u'^2 \, dt$ represent the cost of deviation of the system from a constant state and let $\int_0^T u^2 \, dt$ represent a measure of the quantity of control exerted.†

Let λ, a nonnegative parameter, represent the "price" of control. Then the functional

$$J_\lambda(u) = \int_0^T (u'^2 + \lambda u^2) \, dt \qquad (4.23.3)$$

represents the total cost incurred in the control of the system. As λ varies from 0 to ∞, it is plausible that we use less and less control, which is to

---

† Which we consider the cost of control and to which the cost of deviation is of no importance as far as the analysis is concerned.

say that $\int_0^T u^2\, dt$ decreases monotonically as λ increases, where u is the function that minimizes $J_\lambda(u)$. Consequently, it is reasonable to expect that there is a value of λ for which (4.23.2) holds, provided that k is less than the quantity of control exerted when the cost is zero, that is, when $\lambda = 0$.

It remains to establish all this rigorously, which we now proceed to do. The quantity λ is analogous to the Lagrange parameter of ordinary calculus, and bears the same name in the calculus of variations. In the area of mathematical economics, one can show that it is a price in the ordinary sense of the term.

4.24. Minimizing Property

Suppose that the foregoing assertion is true and that a value, $\bar{\lambda}$, exists for which $\int_0^T \bar{u}^2\, dt = k$. Does $\bar{u}$ minimize $J(u)$ of (4.23.1), subject to the constraint of (4.23.2)? The answer is affirmative and the proof is immediate. Suppose that a function v exists for which

$$J(v) < J(\bar{u}), \qquad \int_0^T v^2\, dt = k, \qquad v(0) = c. \qquad (4.24.1)$$

Then

$$J_\lambda(v) = J(v) + \lambda \int_0^T v^2\, dt = J(v) + \lambda k$$

$$< J(\bar{u}) + \lambda k = J_\lambda(\bar{u}). \qquad (4.24.2)$$

But this contradicts the fact already established that $J_\lambda(u)$, for $\lambda \geq 0$, has an absolute minimum determined by the function $\bar{u}$ satisfying the Euler equation

$$u'' - \lambda u = 0, \qquad u(0) = c, \qquad u'(T) = 0. \qquad (4.24.3)$$

4.25. Monotonicity in λ

The solution of (4.24.3) is, as we know,

$$u = c\, \frac{\cosh \sqrt{\lambda}(T - t)}{\sin \sqrt{\lambda}\, T}. \qquad (4.25.1)$$

It remains to prove that

$$g(\lambda) = c^2 \int_0^T \left(\frac{\cosh \sqrt{\lambda}(T - t)}{\sin \sqrt{\lambda}\, T} \right)^2 dt \qquad (4.25.2)$$

is a strictly monotone decreasing function of λ as λ increases from 0 to ∞. That it is continuous for $\lambda > 0$ is immediate.

A direct approach, using the explicit analytic form of the integrand, does not seem promising. We shall establish this property in this chapter as a consequence of the result concerning the Green's function established above. Subsequently, in a later chapter, we will establish this important result in a much more general setting.

We shall prove the monotonicity of $g(\lambda)$ by showing that u itself considered as a function of λ is monotone decreasing for each t in $[0, T]$. Our aim is to obtain the result as a special case of more general theorems, without any explicit calculations.

4.26. Proof of Monotonicity

Consider the two equations

$$u_1'' - \lambda_1 u_1 = 0, \qquad u_1(0) = c, \quad u_1'(T) = 0,$$
$$u_2'' - \lambda_2 u_2 = 0, \qquad u_2(0) = c, \quad u_2'(T) = 0. \qquad (4.26.1)$$

Our aim is to show that $\lambda_2 > \lambda_1$ implies that $u_1(t) \geq u_2(t)$ for $0 \leq t \leq T$.
Let us write the second equation in the form

$$u_2'' - \lambda_1 u_2 = (\lambda_2 - \lambda_1)u_2, \qquad u_2(0) = c, \qquad u_2'(T) = 0. \qquad (4.26.2)$$

We know that $u_2 > 0$ for $0 \leq t \leq T$. Hence, $(\lambda_2 - \lambda_1)u_2 > 0$. Hence, if we can show that $f(t) \geq 0$, $0 \leq t \leq T$, implies that the solution of

$$u'' - \lambda_1 u = f(t), \qquad u(0) = c, \qquad u'(T) = 0 \qquad (4.26.3)$$

satisfies the inequality $u \leq u_1$, $0 \leq t \leq T$, we will have established the desired result.

One proof is by direct calculation. We have

$$u = u_1 + \int_0^T k(t, t_1)f(t_1)\, dt_1. \qquad (4.26.4)$$

Since we have established the nonpositivity of the Green's function, the proof is complete.

4.27. Discussion

If $\lambda = 0$, the minimum of $\int_0^T u'^2 \, dt$ is attained by $u = c$. The corresponding value of $\int_0^T u^2 \, dt$ is c^2T. Hence, if $k \le c^2T$, we know how to solve the constrained control problem. What if $k > c^2T$?

We could take refuge in the unnaturalness of the problem as far as control theory is concerned. However, the problem is perfectly sensible from the standpoint of Sturm-Liouville theory, which we shall briefly discuss at the end of the chapter. It turns out that we must allow negative prices.

4.28. More General Quadratic Variational Problems

We have previously indicated that a more general form of the kind of variational problem that arises in control theory is that of minimizing the functional

$$J(u, v) = \int_0^T g(u, v) \, dt \qquad (4.28.1)$$

over all u and v connected by the differential equation

$$\frac{du}{dt} = h(u, v), \qquad u(0) = c. \qquad (4.28.2)$$

In order to obtain some ideas as to how to proceed in the general case (ideas we shall put into practice in Volume II), we begin by considering the prototype problem

$$\min_{u,v} \int_0^T (u^2 + v^2) \, dt \qquad (4.28.3)$$

where

$$\frac{du}{dt} = au + v, \qquad u(0) = c. \qquad (4.28.4)$$

Previously, we considered the case $a = 0$.

There are several effective ways in which we can avoid the introduction of any new ideas and methods. To begin with, we can solve for v in (4.28.4) and consider the more familiar problem of minimizing

$$J(u) = \int_0^T (u^2 + (u' - au)^2) \, dt.$$ (4.28.5)

This can be reduced to the form

$$J(u) = \int_0^T [(1 + a^2)u^2 + u'^2] \, dt - 2a \int_0^T u'u \, dt$$

$$= \int_0^T [(1 + a^2)u^2 + u'^2] \, dt - au^2(T) + ac^2,$$ (4.28.6)

a type of functional already discussed.

Alternatively, we can perform the change of variable

$$u = e^{at}w, \qquad v = e^{at}z,$$ (4.28.7)

whence $z = w'$, and replace the original variational problem by that of determining the minimum of

$$\int_0^T e^{2at}(w'^2 + w^2) \, dt,$$ (4.28.8)

where $w(0) = c$.

Each of these methods possesses certain advantages and should by no means be scorned. There is, however, an advantage in considering the problem in its original form, an advantage that becomes considerable when multidimensional, and still more general, control processes are studied.

4.29. Variational Procedure

Following the same general methods used in the discussion of the simpler question, let $\bar{u}$, $\bar{v}$ be a minimizing pair, assumed to exist. Write

$$u = \bar{u} + \varepsilon w, \qquad v = \bar{v} + \varepsilon z,$$ (4.29.1)

where ε is an arbitrary real parameter, $w(0) = 0$, and w and z are otherwise arbitrary in the sense previously described.

Writing

$$J(\bar{u} + \varepsilon w, \bar{v} + \varepsilon z) = J(\bar{u}, \bar{v}) + 2\varepsilon \int_0^T (\bar{u}w + \bar{v}z) \, dt + \varepsilon^2 J(w, z), \quad (4.29.2)$$

we see that the condition that $\bar{u}$, $\bar{v}$ furnish an absolute minimum yields the variational condition

$$\int_0^T (\bar{u}w + \bar{v}z) \, dt = 0 \quad\quad\quad (4.29.3)$$

for all w and z. Since $(\bar{u} + \varepsilon w, \bar{v} + \varepsilon z)$ must satisfy (4.28.4), we have

$$\frac{dw}{dt} = aw + z. \quad\quad\quad (4.29.4)$$

Hence, (4.29.3) yields

$$\int_0^T (\bar{u}w + \bar{v}(w' - aw)) \, dt = 0. \quad\quad\quad (4.29.5)$$

Integrating by parts, this yields the relation

$$[\bar{v}w]_0^T + \int_0^T w(\bar{u} - \bar{v}' - a\bar{v}) \, dt = 0 \quad\quad\quad (4.29.6)$$

for all w. Hence, as above, we suspect that the variational equation is

$$\bar{v}' = -a\bar{v} + \bar{u}, \quad \bar{v}(T) = 0. \quad\quad\quad (4.29.7)$$

Let us henceforth dispense with the bars.

4.30. Proof of Minimum Property

It remains for us to show that the system

$$\begin{aligned} u' &= au + v, & u(0) &= c, \\ v' &= -av + u, & v(T) &= 0, \end{aligned} \quad\quad\quad (4.30.1)$$

has a unique solution and that this solution yields the absolute minimum of $J(u, v)$. Once the existence and uniqueness has been established, the fact that the absolute minimum holds follows from (4.29.2) with $\varepsilon = 1$, as before.

4.31. Existence and Uniqueness

Let us then turn to existence and uniqueness. Let (u_1, v_1) and (u_2, v_2) be principal solutions of

$$u' = au + v,$$
$$v' = -av + u; \tag{4.31.1}$$

that is, $u_1(0) = 1$, $v_1(0) = 0$, $u_2(0) = 0$, $v_2(0) = 1$.

Then the general solution of the linear system has the form

$$u = c_1 u_1 + c_2 u_2,$$
$$v = c_1 v_1 + c_2 v_2, \tag{4.31.2}$$

where c_1 and c_2 are arbitrary constants. The initial condition at $t = 0$ shows that $c_1 = c$. To determine c_2, we employ the condition at $t = T$. This yields the relation

$$0 = v(T) = cv_1(T) + c_2 v_2(T). \tag{4.31.3}$$

Hence, it is necessary to show that $v_2(T) \neq 0$.

We use the same method of proof as before. Suppose that $v_2(T) = 0$. Then, using (4.31.1),

$$u_2{}^2 + v_2{}^2 = u_2(v_2' + av_2) + v_2(u_2' - au_2)$$

$$= u_2 v_2' + u_2' v_2 = \frac{d}{dt}(u_2 v_2). \tag{4.31.4}$$

Hence,

$$\int_0^T (u_2{}^2 + v_2{}^2)\, dt = [u_2 v_2]_0^T = 0, \tag{4.31.5}$$

a contradiction. Thus, $v_2(T) \neq 0$, which shows that (4.30.1) possesses a unique solution.

4.32. The Adjoint Operator

Consider the linear differential operation

$$L(u) = u' - au. \tag{4.32.1}$$

Suppose we ask for the operator $L^*(u)$ with the property that

$$\int_0^T wL(u) \, dt + \int_0^T uL^*(w) \, dt = [\cdots]_0^T, \qquad (4.32.2)$$

that is, that $wL(u) + uL^*(w)$ be an exact differential where L^* is again a linear differential operator. We have

$$\int_0^T w(u' - au) \, dt = [wu]_0^T - \int_0^T u(w' + aw) \, dt. \qquad (4.32.3)$$

Hence, we see that

$$L^*(w) = w' + aw. \qquad (4.32.4)$$

This is called the *adjoint operator*. It is clear from the definition that $(L^*)^* = L$. In other words, the adjoint of the adjoint is the original operator.

The variational equations now take the elegant form

$$L(u) = v, \quad u(0) = c, \quad L^*(v) = u, \quad v(T) = 0. \qquad (4.32.5)$$

This representation is important in connection with duality theory, a subject we shall discuss in a subsequent volume.

EXERCISES

1. Consider

$$J(u) = \int_0^T (u'^2 + \varphi(t)u^2) \, dt.$$

Is it true that

$$\min_{u(0)=c} \, J(u) = \min_{c_2} \left[\min_{\substack{u(0)=c \\ u(T)=c_2}} J(u) \right] ?$$

2. Consider $J(u)$ as above, where $u(0) = c$. Write $u = cu_1 + au_2$, where u_1, u_2 are the principal solutions. Then $J(u) = f(a)$, a quadratic in a. Show that the value of a that minimizes is given by

$$a = -c \, \frac{\int_0^T [u_1'u_2' + \varphi(t)u_1u_2] \, dt}{\int_0^T [u_2'^2 + \varphi(t)u_2^2] \, dt}.$$

Show the equality of the two expressions for the extremal function $u(t)$.

3. Carry through all the details for the minimization of

$$\int_0^T [\varphi_1(t)u'^2 + 2\varphi_2(t)u'u + \varphi_3(t)u^3] \, dt$$

under appropriate assumptions concerning φ_1, φ_2, and φ_3.

4. Show that the solution to the problem of minimizing $\int_0^T (u^2 + v^2) \, dt$ over all v, where $u' = au + v$, $u(0) = c$, is given by

$$u = 2c \left[\frac{a \sinh [(1 + a^2)^{1/2} (T - t)] - (1 + a^2)^{1/2} \cosh [(1 + a^2)^{1/2} (T - t)]}{a \sinh [(1 + a^2)^{1/2} T] - (1 + a^2)^{1/2} \cosh [(1 + a^2)^{1/2} T]} \right].$$

$$v = 2c \left[\frac{\sinh (1 + a^2)^{1/2} (T - t)}{a \sinh [(1 + a^2)^{1/2} T] - (1 + a^2)^{1/2} \cosh [(1 + a^2)^{1/2} T]} \right],$$

$$\min_v \int_0^T (u^2 + v^2) \, dt = \frac{2c^2}{(1 + a^2)^{1/2} \coth [(1 + a^2)^{1/2} T] - a}.$$

Determine the limiting behavior as $T \to \infty$, and as $a \to \infty$.

5. Consider the problem of minimizing $J(u) = \int_0^T (u^2 + u'^2) \, dt$, $u(0) = c$, $u(T) = c_1$. Show that the minimizing function is given by

$$u = \frac{c_1 \sinh t + c \sinh (T - t)}{\sinh T},$$

and

$$\min_u J = (c^2 + c_1{}^2) \coth T - 2cc_1 \coth T.$$

6. Let

$$R(c, T, \lambda) = \min_u \left[\int_0^T (u'^2 + u^2) \, dt + \lambda(u(T) - c)^2 \right], \qquad u(0) = c.$$

Show that the minimizing function is given by

$$u = \frac{\lambda c_1 \sinh T + \lambda c \sinh (T - t) + c \cosh (T - t)}{\lambda \sinh T + \cosh T},$$

$$R(c, T, \lambda) = \frac{\lambda(c^2 + c_1{}^2) \cosh T + c^2 \sinh T - 2\lambda cc_1}{\lambda \sinh T + \cosh T}.$$

7. Consider the problem of maximizing $\int_0^T u^2 \, dt$ subject to the condition $\int_0^T u'^2 \, dt = k_1$, $u(0) = c$. What is the choice of c which maximizes?

4.33. Sturm-Liouville Theory†

We indicated above that the determination of certain critical parameters was intimately connected with Sturm-Liouville theory. Without
going into any details, let us expand on this remark and simultaneously
show where our counterexamples of Section 4.16 came from.

Consider first the time-independent case, where

$$J_\lambda(u) = \int_0^T [u'^2 - \lambda u^2] \, dt. \qquad (4.33.1)$$

Suppose that we want to know when $J_\lambda(u)$ has an absolute minimum over
the class of functions such that $u(0) = u(T) = 0$ and $u' \in L^2[0, T]$. Let us
employ the Fourier expansion

$$u = \sum_{n=1}^\infty a_n \sin \frac{n\pi t}{T},$$

$$u' \sim \sum_{n=1}^\infty \frac{n\pi}{T} a_n \cos \frac{n\pi t}{T}, \qquad (4.33.2)$$

and invoke Parseval's theorem,

$$\int_0^T u^2 \, dt = \frac{T}{2} \sum_{n=1}^\infty a_n{}^2,$$

$$\int_0^T u'^2 \, dt = \frac{T}{2} \sum_{n=1}^\infty \frac{n^2 a_n{}^2 \pi^2}{T^2}. \qquad (4.33.3)$$

The rigorous basis for this is $u' \in L^2(0, T)$. Hence,

$$J_\lambda(u) = \frac{T}{2} \sum_{n=1}^\infty \left(\frac{n^2 \pi^2}{T^2} - \lambda \right) a_n{}^2. \qquad (4.33.4)$$

Thus, if $\lambda < \pi^2/T^2$, we see that $J_\lambda(u) > 0$ unless u is identically zero. On
the other hand, if $\lambda > \pi^2/T^2$, no minimum of $J_\lambda(u)$ exists. If $\lambda = \pi^2/T^2$,
minimum is zero, attained when $u = a_1 \sin \pi t/T$.

† This may be safely skipped by the reader unfamiliar with Sturm-Liouville theory.

The same analysis holds in the more general case where

$$J_\lambda(u) = \int_0^T [u'^2 - \lambda g(t)u^2]\, dt, \qquad (4.33.5)$$

imposing for the sake of simplicity the same conditions $u(0) = u(T) = 0$. The associated Sturm-Liouville equation is

$$u'' + \lambda g(t)u = 0, \qquad u(0) = u(T) = 0. \qquad (4.33.6)$$

Let $\lambda_1 < \lambda_2 < \cdots$ be the characteristic values, and let $u_1(t), u_2(t), \ldots$ be the corresponding characteristic functions, normalized by the condition that $\int_0^T g(t)u_n^2\, dt = 1$.

As is well known, the following identities hold:

(a) $$\int_0^T g u_m u_n\, dt = 0, \qquad m \neq n,$$

(b) $$\int_0^T u_n'^2\, dt = \lambda_n, \qquad (4.33.7)$$

(c) $$\int_0^T u_n' u_m'\, dt = 0, \qquad m \neq n.$$

Hence, if we write

$$u = \sum_{n=1}^\infty a_n u_n,$$

$$u' \sim \sum_{n=1}^\infty a_n u_n', \qquad (4.33.8)$$

we see that

$$J_\lambda(u) = \sum_{n=1}^\infty (\lambda_n - \lambda)a_n^2. \qquad (4.33.9)$$

Hence, $J_\lambda(u) > 0$ for nontrivial u as long as $\lambda < \lambda_1$. The cases $\lambda = \lambda_1$, $\lambda > \lambda_1$ are discussed as before.

4.34. Minimization by Means of Inequalities

Consider the problem of minimizing the positive quadratic functional

$$J(u) = \int_0^T (u'^2 + g(t)u^2)\, dt \qquad (4.34.1)$$

subject to the constraints

$$\int_0^T uf_1\, dt = a_1. \qquad (4.34.2)$$

The approach to problems of this type by means of a Lagrange or Courant parameter can lead to analytic and algebraic complications. Let us then approach the problem using a simple, yet powerful method that can be applied in the study of many more general problems. We will illustrate this in Chapter 7.

The crux of the method is the fact we have been continually exploiting, the positive definite character of J. From this it follows that

$$Q(r_1, r_2) = J(r_1 u + r_2 v) = r_1{}^2 J(u) + r_2{}^2 J(v)$$

$$+ 2r_1 r_2 \int_0^T [u'v' + g(t)uv]\, dt \qquad (4.34.3)$$

is a positive definite quadratic form in the scalar variables r_1 and r_2 for any two functions u and v for which $J(u)$ and $J(v)$ exist. Hence,

$$J(u)J(v) \geq \left(\int_0^T [u'v' + g(t)uv]\, dt \right)^2. \qquad (4.34.4)$$

Integrating by parts in the usual fashion, we see that

$$\int_0^T [u'v' + g(t)uv]\, dt = uv']_0^T + \int_0^T u(g(t)v - v'')\, dt$$

$$= \int_0^T uE(v)\, dt, \qquad (4.34.5)$$

where we write $E(v) = g(t)v - v''$, provided that we impose the conditions

$$v'(0) = v'(T) = 0. \qquad (4.34.6)$$

Thus, (4.34.4) yields

$$J(u)J(v) \geq \left(\int_0^T uE(v) \, dt \right)^2. \tag{4.34.7}$$

Set

$$E(v) = f_1, \qquad v'(0) = v'(T) = 0, \tag{4.34.8}$$

yielding

$$v = - \int_0^T K(t, t_1) f_1(t_1) \, dt_1 \tag{4.34.9}$$

in terms of the appropriate Green's function. Then (4.34.7) reads

$$J(u)J\left(\int_0^T K(t, t_1) f_1(t_1) \, dt_1 \right) \geq \left(\int_0^T u f_1 \, dt \right)^2. \tag{4.34.10}$$

Hence,

$$\min_u J(u) = \frac{a_1^2}{J(\int_0^T K(t, t_1) f_1(t_1) \, dt_1)}. \tag{4.34.11}$$

Retracing our steps, we see that equality holds in (4.34.4) if and only if $r_1 u + r_2 v = 0$ for some r_1 and r_2; that is, if $u = b_1 v$ for some constant b_1. Thus, the minimum is assumed for

$$u = b_1 \int_0^T K(t, t_1) f_1(t_1) \, dt_1. \tag{4.34.12}$$

This constant b_1 is determined by the condition of (4.34.2). Thus,

$$b_1 \int_0^T \int_0^T K(t, t_1) f_1(t) f_1(t_1) \, dt \, dt_1 = a_1. \tag{4.34.13}$$

Since $K(t, t_1)$ is a positive definite kernel, the double integral is nonzero and thus b_1 is uniquely determined.

4.35. Multiple Constraints

Suppose there are a number of constraints of the foregoing type,

$$\int_0^T u f_i \, dt = a_i, \qquad i = 1, 2, \ldots, M. \tag{4.35.1}$$

A simple extension of the foregoing method yields the desired minimum value.

We begin with the quadratic form in three variables,

$$Q(r_1, r_2, r_3) = r_1{}^2 J(u) + r_2{}^2 J(v) + r_3{}^2 J(w)$$

$$+ 2r_1 r_2 \int_0^T [u'v' + a(t)uv]\, dt + \cdots, \qquad (4.35.2)$$

and the observation that it is positive definite. From this, we conclude the inequality

$$\begin{vmatrix} J(u) & \int_0^T [u'v' + g(t)uv]\, dt & \int_0^T [u'w' + g(t)uw]\, dt \\[2mm] \int_0^T [\cdots]\, dt & J(v) & \int_0^T [\cdots]\, dt \\[2mm] \int_0^T [\cdots]\, dt & \int_0^T [\cdots]\, dt & J(w) \end{vmatrix} \geq 0.$$

$$(4.35.3)$$

Integrating the off-diagonal terms by parts as before, this yields the inequality

$$\begin{vmatrix} J(u) & \int_0^T uE(v)\, dt & \int_0^T uE(w)\, dt \\[2mm] \int_0^T uE(v)\, dt & J(v) & \int_0^T vE(w)\, dt \\[2mm] \int_0^T uE(w)\, dt & \int_0^T vE(w)\, dt & J(w) \end{vmatrix} \geq 0. \qquad (4.35.4)$$

Proceeding as before, we readily determine both the minimizing function and the minimum value of $J(u)$.

EXERCISES

1. Minimize $\int_0^T u'^2\, dt$ subject to $\int_0^T u\, dt = 1$, and then subject to the additional constraint $\int_0^T tu\, dt = 0$.
2. Determine the minimum of $J(u)$ subject to $\int_0^T u f_1\, dt_1 = a_1$, $u(t_2) = a_2$, where $0 \leq t_2 \leq T$, by considering the point condition to be the limit

of a condition of the form $\int_0^T u f_2 \, dt_1 = a_2$ as f_2 approaches a delta function.

3. What is the maximum of $u(t_2)$ subject to the condition $\int_0^T u'^2 \, dt = 1$? What is the maximum value of $u(t_2)$ subject to $\int_0^T (u'^2 + g(t)u^2) \, dt = 1$, under the assumption that the quadratic functional is positive definite?

4.36. Unknown External Forces

So far we have considered the problem of the control of a system where cause and effect are completely known. How do we formulate the question of control of a system subject to some unknown influences? One approach is to use the theory of probability, as we shall do in Volume III. Another approach, basically pessimistic and indeed almost paranoiac, is to suppose that the unknown effects combine to produce the most undesirable results.

An analytic formulation based on these ideas is the following. Let

$$\frac{du}{dt} = au + v + w, \qquad u(0) = c, \qquad (4.36.1)$$

and suppose that v is chosen to minimize

$$J(u, v, w) = \int_0^T (u^2 + v^2 - w^2) \, dt, \qquad (4.36.2)$$

and that w is chosen to maximize.

A first question is that of the order of these operations. Do we want to determine

$$\max_w \min_v J(u, v, w), \qquad (4.36.3)$$

or, alternatively,

$$\min_v \max_w J(u, v, w) ? \qquad (4.36.4)$$

The first corresponds to a choice of $w(t)$ before $v(t)$, an obvious disadvantage.

Rather remarkably, it turns out that in this case, and in many others of importance, the two expressions are equal. We leave to the reader the verification of this by direct calculation using the foregoing results.

A direct proof without calculation follows from basic results in the mathematical theory of games.

Miscellaneous Exercises

1. Show that $u(x) + \int_a^b k(x, y)u(y)\, dy = f(x)$ is the variational equation associated with the functional

$$J(u) = \int_a^b u^2\, dx + \int_a^b \int_a^c k(x, y)u(x)u(y)\, dx\, dy - 2\int_a^b f(x)u(x)\, dx$$

   if $k(x, y)$ is symmetric, that is, $k(x, y) = k(y, x)$.

2. What is the minimum of $\int_0^T u'^2\, dt$ subject to the condition $u(0) = c$, $\int_0^T u^2\, dt \le k$?

3. Consider the problem of minimizing the function

$$J(u, \lambda_1, \lambda_2) = \int_0^T (u^2 + v^2)\, dt + \lambda_1 \int_0^T (u' - au - v)^2\, dt$$
$$+ \lambda_2(u(0) - c)^2$$

   where $\lambda_1, \lambda_2 > 0$, and u, v are now independent functions. Let $\bar{u} = \bar{u}(t, \lambda_1, \lambda_2)$. What happens $\lambda_1 \to \infty$, $\lambda_2 \to \infty$ independently or jointly?

4. Consider the problem of minimizing

$$J_\varepsilon(u) = \int_0^T [\varepsilon u'^2 + g(t)u^2]\, dt,$$

   where $g(t) \ge 0$ and continuous for $0 \le t \le T$. What is $\lim_{\varepsilon \to 0} J_\varepsilon(u)$, where u is constrained by $u(0) = 1$? If $u(t, \varepsilon)$ denotes the function that minimizes, does $\lim_{\varepsilon \to 0} u(t, \varepsilon)$ exist? Consider first the case $g(t) = 1$ where the calculation can be done explicitly.

5. Show that the condition that $u \in L^2(0, \infty)$ where $u'' - u = h$ with $h \in L^2(0, \infty)$ implies that

$$u = c_2\, e^{-t} + \frac{e^t}{2} \int_t^\infty e^{-s} h(s)\, ds - \frac{e^{-t}}{2} \int_0^t e^s h(s)\, ds$$

   for some constant c_2.

6. Hence, show that if $u, u'' \in L^2(0, \infty)$, then $u' \in L^2(0, \infty)$. (*Hint:* Write $u'' - u = h$, where $h = u'' - u$.)
7. Establish corresponding results for the case where $|u|, |u''|$ are uniformly bounded for $t \geq 0$.
8. Using the identity $(d/dt)[e^{-t}(u'' + u)] = e^{-t}(u'' - u)$, show that

$$u' = -u - e^t \int_t^\infty e^{-s}[u'' - u] \, ds.$$

9. Hence, show that

$$\max_{t \geq 0} |u'| \leq 2 \max_{t \geq 0} |u| + \max_{t \geq 0} |u''|.$$

10. Replace $u(t)$ by $u(rt)$ where $r > 0$ and deduce that

$$r \max_{t \geq 0} |u| \leq 2 \max_{t \geq 0} |u| + r^2 \max_{t \geq 0} |u''|$$

for any $r > 0$. Hence, conclude that

$$\left(\max_{t \geq 0} |u'| \right)^2 \leq 8 \left(\max_{t \geq 0} |u| \right) \left(\max_{t \geq 0} |u''| \right).$$

11. Obtain corresponding inequalities connecting $\int_0^\infty u^2 \, dt$, $\int_0^\infty u'^2 \, dt$, and $\int_0^\infty u''^2 \, dt$. (The constant 8 is not the best possible constant. For further results along these lines, see

G. H. Hardy, J. E. Littlewood, and G. Polya, *Inequalities*;
E. F. Beckenbach and R. Bellman, *Inequalities*, Springer, Berlin, 1961,

where many further references may be found.)
12. Show that if $(u'' + a_1(t)u' + a_2(t)u) \in L^2(0, \infty)$, $u \in L^2(0, \infty)$ and $|a_1| \, |a_2| \leq c_1 < \infty$ for $t \geq 0$, then $u' \in L^2(0, \infty)$.
13. Show that

$$\int_0^a u(u'' - qu) \, dt = [uu']_0^a - \int_0^a (u'^2 + qu^2) \, dt.$$

14. Hence show that $u'' - qu \geq 0$, $0 \leq t \leq a$, $u(0) = u(a) = 0$, $\int_0^a (u'^2 + qu^2) \, dt \geq 0$ imply that $u \leq 0$ for $0 \leq t \leq a$.
15. Consider the linear differential operator $L(u) = u'' + a_1(t)u' + a_2(t)u$. Write

$$W_0 = 1, \qquad W_1 = u_1, \qquad W_2 = \begin{vmatrix} u_1 & u_2 \\ u_1' & u_2' \end{vmatrix}.$$

Show that in $0 \leq t \leq a$

$$L(u) = \frac{W_2}{W_1} \frac{d}{dt} \left[\frac{W_1{}^2}{W_2} \frac{d}{dt} \left[\frac{u}{u_1} \right] \right],$$

provided that u_1 and u_2 are two functions such that W_1 and W_2 are positive in $[0, a]$.

16. Let $v(t)$ be a real function satisfying the Riccati equation $v' + v^2 + p(t) = 0$ with an appropriate condition at $t = 0$. Suppose that v exists for $0 \leq t \leq 2\pi$ and is nonnegative there. Show that

$$\int_0^{2\pi} [u' - v(u - b)]^2 \, dt + v(2\pi)[u(2\pi) - b]^2$$

$$= \int_0^{2\pi} [u'^2 - p(u - b)^2] \, dt.$$

17. What sufficient conditions can one deduce from this that $\int_0^{2\pi} u'^2 \, dt \geq \int_0^{2\pi} pu^2 \, dt$?

18. Establish Wirtinger's inequality: If $u(t)$ has period 2π and $\int_0^{2\pi} u \, dt = 0$, then $\int_0^{2\pi} u'^2 \, dt \geq \int_0^{2\pi} u^2 \, dt$ with strict inequality unless $u = c_1 \cos t + c_2 \sin t$.
(See

G. H. Hardy, J. E. Littlewood, and G. Polya, *Inequalities*, Cambridge, London, 1934;

E. F. Beckenbach and R. Bellman, *Inequalities*, Springer, Berlin, 1961;

P. R. Beesack, "Integral Inequalities of the Wirtinger Type," *Duke Math. J.*, **25**, 1948, pp. 477–498;

W. J. Coles, "A General Wirtinger-Type Inequality," *Duke Math. J.*, **27**, 1960, pp. 133–138.)

19. Show that

$$\left(\int_0^T f^2 \, dt \right)^{1/2} = \max_g \int_0^T fg \, dt,$$

where the maximum is taken over all g such that $\int_0^T g^2 \, dt = 1$.

20. Hence, establish the Minkowski inequality,

$$\left(\int_0^T (f + h)^2 \, dt \right)^{1/2} \leq \left(\int_0^T f^2 \, dt \right)^{1/2} + \left(\int_0^T h^2 \, dt \right)^{1/2}.$$

When does equality hold?

21. Consider the equation $u'' + a(t)u = 0$, $u(0) = c_1$, $u(T) = c_2$. Let a_1 be an initial approximation to $u'(0)$ and let $u_1(t)$ be determined as the solution of $u_1'' + a(t)u_1 = 0$, $u_1(0) = c_1$, $u_1'(0) = a_1$. Let $b_1 = u_1'(T)$ and u_2 be determined as the solution of $u'' + a(t)u = 0$, $u(T) = c_2$, $u'(T) = b_1$, and $a_2 = u_2'(0)$. Let u_3 be determined as the solution of $u_3'' + a(t)u_3 = 0$, $u_3(0) = c_1$, $u_3'(0) = a_2$, etc. Under what conditions does the sequence $\{u_n\}$ converge? (See

R. Bellman, "On the Iterative Solution of Two-Point Boundary Value Problems," *Boll. Un. Mat. Ital.*, **16**, pp. 145–149.
T. A. Brown, *Numerical Solution of Two-Point Boundary Value Problems for Linear Differential Equations*, 1963, unpublished.)

22. Consider the functional $J(u) = \int_0^T (\varepsilon u''^2 + u'^2 + u^2)\, dt$, where $\varepsilon > 0$, and $u(0) = c_1$, $u'(0) = c_2$. Let $u(t, \varepsilon)$ denote the function which minimizes $J(u)$. Then
    (a) As $\varepsilon \to 0$, does $\lim J(u)$ exist?
    (b) As $\varepsilon \to 0$, does $\lim u(t, \varepsilon)$ exist?
    (c) What is the relation, if any, between $\lim_{\varepsilon \to 0} J(u)$ and the minimum of $J_1(u) = \int_0^T (u'^2 + u^2)\, dt$?

23. Assume that all of the following integrals exist. Then

$$\int_0^\infty (u'' + u)^2\, dt = \int_0^\infty u''^2\, dt + \int_0^\infty u^2\, dt - 2\int_0^\infty u'^2\, dt,$$

provided that either $u(0)$ or $u'(0)$ is zero. Hence, for any positive r,

$$r^4 \int_0^\infty u''^2\, dt + \int_0^\infty u^2\, dt - 2r^2 \int_0^\infty u'^2\, dt \geq 0,$$

and thus

$$\left(\int_0^\infty u''^2\, dt\right)\left(\int_0^\infty u^2\, dt\right) \geq \left(\int_0^\infty u'^2\, dt\right)^2,$$

if either $u(0) = 0$ or $u'(0) = 0$.

24. Determine the minimum of $\int_0^\infty (u^2 + b_1 v^2)\, dt$ over v where $au'' + bu' + cu = v$, $u(0) = c_1$, $u'(0) = c_2$. (See

O. L. R. Jacobs, "The Damping Ratio of an Optimal Control System," *IEEE Trans. Auto. Control*, **AC-10**, 1965, pp. 473–476.)

25. Determine the minimum over v of $\int_0^\infty (u^2 + b_2(u'')^2)\, dt$. (See

G. C. Newton, L. A. Gould, and J. F. Kaiser, *Analytic Design of Linear Feedback Controls*, Wiley, New York, 1957, Chapter 2.)

26. Determine the minimum over v of $\int_0^\infty (u^2 + u'^2)\, dt$. (See

R. E. Kalman and J. E. Bertram, "Control System Analysis and Design via the Second Method of Lyapunov," *Trans. ASME*, Ser. D, **82**, 1960, pp. 371–393.

C. W. Merriam III, *Optimization Theory and the Design of Feedback Control Systems*, McGraw-Hill, New York, 1964.)

27. If one assumes in all these cases that the optimal control is linear, $v = b_3 u + b_4 u'$, where b_3 and b_4 are constants, how does one determine these constants?

28. Show that the problem of minimizing

$$J(u) = \frac{\int_0^T u'^2\, dt}{\int_0^T g(t)u^2\, dt}$$

where $g(t) > 0$, $\int_0^T g(t)u^2\, dt = 1$, $u(0) = (T) = 0$, leads to the Sturm-Liouville equation $u'' + \lambda g(t)u = 0$, $u(0) = u(T) = 0$.

29. Show directly from this that if $q(t) \geq p(t) > 0$, then the first zero of the solution of $v'' + q(t)v = 0$, $v(0) = 0$, $v'(0) = 1$, cannot occur after the first zero of the solution of $u'' + p(t)u = 0$, $u(0) = 0$, $u'(0) = 1$.

30. Show directly from this that the first characteristic function preserves one sign inside of $(0, T)$.

31. Show directly from the variational characterization that the first characteristic function of $u'' + \lambda p(t)u = 0$, $u(0) = u(T) = 0$, where $p(t) > 0$, is unimodal. (*Hint:* Suppose $u(t)$ has the form

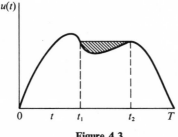

Figure 4.3

and consider what happens if we redefine $u(t) = u(t_2)$ for $t_1 \le t \le t_2$. See

R. Bellman, "Oscillatory and Unimodal Properties of Solutions of Second-Order Linear Differential Equations," *Boll. Un. Mat. Ital.,* **19**, 1964, pp. 306–310.

The point of the foregoing methods is that they generalize to higher-dimensional systems and to partial differential equations.)

32. Consider the equation $u'' - u - 2u^3 = 0$, $u(0) = 0$, $u'(0) = c_1 > 0$. Show that u is nonzero for $t > 0$ in two ways:
 (a) Assume that $u(T) = 0$ and consider $\int_0^T u(u'' - u - 2u^3)\, dt$.
 (b) Assume that T is the first value for which $u(T) = 0$ and examine the consequence of the fact that u has at least one relative maximum in $[0, T]$.

33. Consider the equation $u'' - u - 2u^3 = 0$, $u(0) = 0$, $u'(T) = 0$. Show that there is at most one solution by using the fact that the difference of two solutions, u, v, satisfies the equation $(u - v)'' - (u - v) - 2(u - v)(u^2 + uv + v^2) = 0$ and the nonnegativity of u and v.

34. Show that the solution of $u'' - u - 2u^3 = 0$, $u(0) = 0$, $u'(0) = c$, exists for $t \ge 0$ for any c. Show that $u'(T)$ is a continuous function of c that assumes both positive and negative values as c ranges over $(-\infty, \infty)$, and thus that there is at least one value of c for which $u'(T) = 0$.

35. Show that u_c satisfies the equation $v'' - v(1 + 6u^2) = 0$, $v(0) = 0$, $v'(0) = 1$, and thus that u is a monotone increasing function of c.

36. Generalize the foregoing results to cover the equation $u'' - g(u) = 0$, $u(0) = 0$, $u'(T) = 0$, under appropriate assumptions concerning g.

37. Suppose that we employ a perturbation procedure to minimize

$$J(u, \varepsilon) = \int_0^T (u'^2 + u^2 + \varepsilon u^4)\, dt,$$

where $\varepsilon > 0$, and $u(0) = c$. Show that the Euler equation is $u'' - u - 2\varepsilon u^3 = 0$, $u(0) = c$, $u'(T) = 0$. Set $u = u_0 + \varepsilon u_1 + \cdots$, where $u_0, u_1, \ldots$, are independent of T. Then

$$u_0'' - u_0 = 0, \qquad u_0(0) = c, \quad u_0'(T) = 0,$$
$$u_1'' - u_1 - 2u_0^3 = 0, \qquad u_1(0) = 0, \quad u_1'(T) = 0,$$

and so on. Show that $u_0, u_1, \ldots,$ are uniquely determined, and exhibit u_0, u_1 explicitly.

38. Does the perturbation series for the solution of the Euler equation converge for all $\varepsilon \geq 0$? If not, why not? Does the functional $J(u, \varepsilon)$ possess an absolute minimum for all $\varepsilon \geq 0$?

39. Show that

$$J(u, \varepsilon) = J(u_0, \varepsilon) + \varepsilon^2 \left[J(u_1, 0) + 4 \int_0^T u_0{}^3 u_1 \, dt_1 \right] + \cdots,$$

and thus that u_1 can be determined by the requirement that it minimize $J(u_1, 0) + 4 \int_0^T u_0{}^3 u_1 \, dt_1$ subject to $u_1(0) = 0$. Are all functions $u_2, u_3, \ldots$, equally determined as the solution of quadratic variational problems? (*Hint:* Use Exercise 37 to answer this.)

40. Use the relation $u^2 = \max_v (2uv - v^2)$ to show that

$$J(u) = \int_0^T (u'^2 + \varphi(t)u^2) \, dt = \max_v \left[\int_0^T (u'^2 + \varphi(t)(2uv - v^2)) \right] dt,$$

if $\varphi(t) \geq 0$. Hence, show that

$$\min_u J(u) = \min_u \max_v \left[\int_0^T (u'^2 + \varphi(t)(2uv - v^2)) \, dt \right].$$

The minimum is over functions of u for which $u(0) = c$.

41. Introduce the functional

$$F(v) = \min_u \left[\int_0^T (u'^2 + \varphi(t)(2uv - v^2)) \right] dt,$$

and calculate it as a functional of v.

42. Determine the maximum of $F(v)$ over all v such that $F(v)$ exists and for which $v(0) = c$.

43. On the basis of this explicit calculation, show that

$$\min_u \max_v \left[\int_0^T (u'^2 + \varphi(t)(2uv - v^2)) \, dt \right] = \max_v \min_u [\cdots],$$

and thus that $\min_u J(u) = \max_v F(v)$.

44. On the basis of this last result, show how to obtain upper and lower bounds for $\min_u J(u)$, using simple trial functions. In particular, what happens if we use the solution for the infinite interval, $u = ce^{-t}$?

45. Use the relation $u'^2 = \max_v (2u'v - v^2)$ to show that

$$\min_u \int_0^T (u'^2 + \varphi(t)u^2) \, dt = \min_u \max_v \left[\int_0^T (2u'v - v^2 + \varphi(t)u^2) \, dt \right].$$

Suppose that $\varphi(t) \geq \delta > 0$.

46. Integrate by parts to show that

$$\min_u J(u) = \min_u \max_v \left[-2cv(0) + \int_0^T [\varphi(t)u^2 - v^2 - 2uv'] \, dt \right].$$

Why can we safely assume that $v(T) = 0$?

47. Calculate

$$F(v) = \min_u \left[-2cv(0) + \int_0^T [\varphi(t)u^2 - v^2 - 2uv'] \, dt \right].$$

Show that the value of u that minimizes is $\bar{u} = v'/\varphi(t)$. Hence,

$$\max_v F(v) = \max_v \left[-\int_0^T \left[\frac{v'^2}{\varphi(t)} + v^2 - 2cv' \right] dt \right].$$

The Euler equation is

$$\frac{d}{dt}\left(\frac{v'}{\varphi(t)}\right) - v = 0, \qquad v(T) = 0, \qquad \left(\frac{v'}{\varphi(t)} - c\right)_{t=0} = 0.$$

48. Calculate the value of $\max_v F(v)$ and thus show directly that $\max_v F(v) = \min_u J(u)$.

(For the foregoing, see

R. Bellman, "Quasilinearization and Upper and Lower Bounds for Variational Problems," *Quart. Appl. Math.*, **19**, 1962, pp. 349–350.)

49. Use the trial function $u = c + a_1 t + a_2 t^2$ with appropriate choice of a_1, a_2 to get upper and lower bounds for $J(u)$ for $T \ll 1$; use the trial function $u = ce^{-t} + a_1 e^{-bt}$ to get upper and lower bounds for $T \gg 1$.

50. Extend the foregoing results to the case where

$$J(u) = \int_0^T (u'^2 + g(u)) \, dt,$$

using first the representation $u'^2 = \max_v (2u'v - v^2)$, and next the result

$$g(u) = \max_v [g(v) + (u - v)g'(v)].$$

(The basic idea underlying the foregoing results is that of duality. We will pursue this in great detail in one of the later volumes. Meanwhile, the reader may wish to refer to

R. Bellman, "Functional Equations and Successive Approximations in Linear and Nonlinear Programming," *Naval Res. Logist. Quart.*, **7**, 1960, pp. 63–83.
R. Bellman, "Quasilinearization and Upper and Lower Bounds for Variational Problems," *Quart. Appl. Math.*, **19**, 1962, pp. 349–350.
J. D. Pearson, "Reciprocity and Duality in Control Programming Problems," *J. Math. Anal. Appl.*, **10**, 1965, pp. 388–408,

where references to earlier work by Friedrichs, Trefftz, and others may be found.)

BIBLIOGRAPHY AND COMMENTS

One of our aims in this chapter has been to emphasize to the reader that the solution of a variational problem is to be considered to be a function not only of time, but also of the initial condition, the duration of the process, and of various parameters that appear in the differential equation and the quadratic functional.

4.2. In Chapter 9, we discuss a direct existence proof using functional analysis. For a detailed discussion of inequalities, see

E. F. Beckenbach and R. Bellman, *An Introduction to Inequalities*, Random House, New York, 1961.
E. F. Beckenbach and R. Bellman, *Inequalities*, Springer, Berlin, 1961.

4.3. Euler derived his equation in a different fashion, using a passage to the limit from the case of a finite-dimensional variational problem.

4.5. From the standpoint of control theory, Haar's approach is a natural one since z', the rate of change of state, is the more fundamental variable than z, the state.

4.9. For further results in asymptotic control theory, see

R. Bellman and R. Bucy, "Asymptotic Control Theory," *SIAM Control*, **2**, 1964, pp. 11–18.
R. Bellman, "Functional Equations in the Theory of Dynamic Programming—XIII: Stability Considerations," *J. Math. Anal. Appl.*, **12**, 1965, pp. 537–540.

4.10. In the exercises we are hinting at the use of the Rayleigh-Ritz method. For detailed discussion and applications, see

R. Bellman and J. M. Richardson, *Introduction to Methods of Nonlinear Analysis*, To appear.

4.15. For a discussion of successive approximations via quasilinearization, see

R. Bellman and R. Kalaba, *Quasilinearization and Nonlinear Boundary Value Problems*, American Elsevier, New York, 1965.

4.21.–4.22. For further discussion of the results given here and in the exercises, see

R. Bellman, "On the Nonnegativity of Green's Functions," *Boll. Un. Mat. Ital.*, **12**, 1957, pp. 411–413.
R. Bellman, "On Variation-Diminishing Properties of Green's Functions," *Boll. Un. Mat. Ital.*, **16**, 1961, pp. 164–166.
R. Bellman, "On the Nonnegativity of Green's Functions," *Boll. Un. Mat. Ital.*, **18**, 1963, pp. 219–221.

4.23–4.24. See

R. Bellman and S. Dreyfus, *Applied Dynamic Programming*, Princeton Univ. Press, Princeton, New Jersey, 1962,
for a further discussion.

4.33. For a detailed discussion of Sturm-Liouville theory, see

E. L. Ince, *Ordinary Differential Equations*, Dover, New York, 1944.

4.34. For the background in matrix theory and quadratic forms, see

R. Bellman, *Introduction to Matrix Analysis*, McGraw-Hill, New York, 1960.

4.36. For basic ideas in the theory of games, see

J. von Neumann and O. Morgenstern, *The Theory of Games and Economic Behavior*, Princeton Univ. Press, Princeton, New Jersey, 1948.

5

DYNAMIC PROGRAMMING

5.1. Introduction

In this chapter, we will introduce the reader to an entirely different approach to the treatment of control processes. It will be based on the theory of dynamic programming, a mathematical abstraction and extension of the fundamental engineering concept of feedback control.

Our initial aim is to make the concepts and methodology as apparent and plausible as possible. Consequently, we shall at first proceed in a formal fashion, and then use the preceding results to establish the validity of the formulas we obtain for the cases we consider.

The last part of the chapter will be devoted to the topic of discrete control processes.

5.2. Control as a Multistage Decision Process

Consider the problem of minimizing the functional

$$J(u) = \int_0^T (u'^2 + u^2) \, dt \qquad (5.2.1)$$

over all $u(t)$ such that $u' \in L^2[0, T]$ and subject to the initial condition $u(0) = c$, a problem we completely resolved in the preceding chapter. Our approach was a straightforward extension of the ideas of calculus. The existence of a minimizing function $\bar{u}$ was assumed and the variation of $J(u)$ in the neighborhood of this extremum provided a necessary condition, the Euler equation, which we showed was also sufficient.

We now want to employ some different techniques generated by the idea that the minimization problem can be considered to arise from a control process. In place of asking for the solution u as a function of time, we ask for the most efficient control at each point in phase space. At any particular point in the history of the process, we want instructions as to what to do next.

In terms of the foregoing minimization question, at the point $P = P(u, t)$, we want to determine u' as a function of u and t. See Fig. 5.1. Intuitively, we conceive of this as a guidance process in which

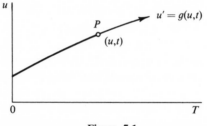

Figure 5.1

it is required to furnish steering directions continuously.† A control process is thus considered to be a multistage decision process consisting of the following operations:

(a) At time t, the state of the system is observed.
(b) Based on this information and a control law, a decision is made.   (5.2.2)
(c) This decision produces an effect upon the system.

In order to keep the initial presentation simple, in this volume we assume, quite unrealistically of course, that it is always possible to

---

† Let us note in passing that the word for steersman in Greek has given rise to the currently fashionable term "cybernetics."

observe the system and obtain complete information about its state, that no time or effort is required for this, that the correct decision is always made, that we know exactly what the effect of every decision is, and, finally, that we know what the purpose of the control process is.

The mention of these more realistic aspects of control is made to give the reader some brief idea of the scope of control theory and of the vast amount of research that remains to be done. In particular, we wish to emphasize the grave difficulties entailed in fitting real control processes into simplified molds. Nevertheless, there is considerable merit in considering the simpler versions in detail, and, remarkably, on occasion the case even occurs that these naive formulations are useful.

5.3. Preliminary Concepts

As indicated above, the direction of the path at a particular point P is to be considered dependent upon the coordinates of the point, the state u and the time, as measured in terms of time transpired or time remaining in the process. This point of view combined with the additive nature of the quadratic functional used to evaluate the control process enables us to regard each decision that is made as the initial decision of a new control process. Consequently, we can constrain our attention to the initial slope, the value $u'(0)$, and regard this as a function of c, the initial state, and T, the duration of the process. This basic function, which we denote by $v(c, T)$, is called a *policy*. A policy is a rule for prescribing the action to take at every possible position of the system in phase space. The term is used deliberately because it agrees so closely with the intuitive concept of " policy."

The analytic structure of the policy function depends upon the way in which the system has been described in mathematical terms. There are many ways of constructing a mathematical formulation and it is not always clear what is the irreducible amount of information concerning the system that is required for either optimal or effective control. Careful analysis of this type is essential in the study of large complex systems occurring in the economic, industrial, and biomedical area. It is important to emphasize that the concepts of " state " and " policy " are elusive and require continuing examination.

A policy that minimizes the functional representing the overall cost of the control process, or, generally, that most effectively achieves the desired goal, is called an *optimal policy*. Even though there is a unique minimum or maximum value, there need not be a unique optimal policy that attains this value. There may be several ways of carrying this out.

As we shall see, there is a very simple and intuitive characterization of optimal policies, the principle of optimality. This principle can be used to obtain functional equations that simultaneously determine the required minimum or maximum and all optimal policies.

It is of some interest to point out that the concept of policy is more fundamental by far than that of minimizing function and thus that dynamic programming can profitably be applied in many situations where there is either no criterion function or a vector-valued criterion function. This is particularly the case in simulation.

5.4. Formalism

We begin with the observation that the minimum value of $J(u)$ depends on the two quantities singled out above: the initial state c and the duration of the process T. Hence, following the usual mathematical translation of "depends on," we may write

$$f(c, T) = \min_u J(u). \tag{5.4.1}$$

This function is defined for $T \geq 0$ and $-\infty < c < \infty$. The original minimization problem is thus imbedded within a family of variational problems of the same type. It remains to show that this is a useful step to take in our pursuit of the solution of the original problem.

This dependence is intuitively clear, and, of course, analytically clear by virtue of the complete solution of the minimization problem obtained in the previous chapter. To obtain an equation for $f(c, T)$, which, as it turns out, will provide the optimal policy for us, we argue as follows (see Fig. 5.2). Let Δ be an infinitesimal, and let $v = v(c, T)$ denote the initial slope at $t = 0$. Then at time Δ, the new state will be $c + v\Delta$, to terms in Δ^2.

We have

$$J(u) = \int_0^\Delta + \int_\Delta^T .\qquad (5.4.2)$$

It is this additivity of the functional representing the total cost, together with our assumptions of instantaneous observation, decision, and action, that enables us to regard the problem of determining the form of the minimizing function over (Δ, T) as a new control process starting in state $c + v\Delta$ with duration $T - \Delta$. Hence, we maintain on the basis of common sense (and with prior knowledge that we have a proof by

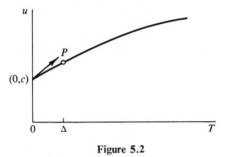

Figure 5.2

contradiction at our disposal) that regardless of how v has been chosen at $t = 0$, we will continue from P (see Fig. 5.2) at time Δ in such a way as to minimize the cost incurred,

$$\int_\Delta^T (u'^2 + u^2)\, dt,\qquad (5.4.3)$$

over the remaining time interval (Δ, T). By definition of the function $f(c, T)$, this minimum value is $f(c + v\Delta, T - \Delta)$.

We are considering Δ to be an infinitesimal† and systematically neglecting terms of order Δ^2. Hence, we can write

$$\int_0^\Delta (u'^2 + u^2)\, dt = (v^2 + c^2)\Delta + O(\Delta^2),\qquad (5.4.4)$$

---

† A shorthand notation for a term that is initially small and ultimately will be made to tend to zero.

as an estimate of the cost of control over the initial interval $[0, \Delta]$. Thus, (5.4.2) takes the form

$$J(u) = (v^2 + c^2)\Delta + f(c + v\Delta, T - \Delta) + O(\Delta^2). \qquad (5.4.5)$$

How should v be chosen? It is reasonable to suppose that we choose v to minimize the right-hand side. We must balance the cost of control over $[0, \Delta]$ against the cost of optimal control over $[\Delta, T]$. Hence, we have

$$f(c, T) = \min_v \left[(v^2 + c^2)\Delta + f(c + v\Delta, T - \Delta) + O(\Delta^2) \right]. \qquad (5.4.6)$$

There are serious (but not fatal) difficulties in the way of a rigorous procedure along these lines, since the term $O(\Delta^2)$ is dependent on v. However, proceeding as before, we first push the formalism to its illogical limit, and then worry about the interpretation and validation of the results obtained in this cavalier fashion.

Expanding in a Taylor series, we have

$$f(c + v\Delta, T - \Delta) = f(c, T) + v\frac{\partial f}{\partial c}\Delta - \frac{\partial f}{\partial T}\Delta + O(\Delta^2). \qquad (5.4.7)$$

Observe that we assume in doing this that f possesses partial derivatives with respect to c and T that are sufficiently smooth. At the moment we have no justification for this. But two illegalities in mathematics are no worse than one! Hence, (5.4.6) becomes

$$f(c, T) = \min_v \left[(v^2 + c^2)\Delta + f(c, T) + v\frac{\partial f}{\partial c}\Delta - \frac{\partial f}{\partial T}\Delta + O(\Delta^2) \right].$$
$$(5.4.8)$$

Cancelling the common terms $f(c, T)$ on both sides, dividing through by Δ, and letting $\Delta \to 0$, we obtain the nonlinear partial differential equation

$$\frac{\partial f}{\partial T} = \min_v \left[c^2 + v^2 + v\frac{\partial f}{\partial c} \right]. \qquad (5.4.9)$$

The initial condition is clearly $f(c, 0) = 0$. As we shall see, this equation will provide us with both the minimum value $f(c, T)$ and the optimal policy $v(c, T)$.

1. Use the foregoing method to obtain equations for

$$f_1(c, T) = \min_{u(0)=c} \left[\int_0^T (u'^2 + u^2) \, dt + \lambda u(T)^2 \right],$$

$$f_2(c, T) = \min_{u(0)=c} \left[\int_0^T (u'^2 + u^2) \, dt + \lambda (u(T) - a)^2 \right].$$

Observe the difference in the initial conditions at $T = 0$.

2. Obtain a rigorous derivation of the formula in (5.4.9), starting from (5.4.6), by use of the knowledge that $f(c, T) = c^2 r(T)$.

5.5. Principle of Optimality

The argument we used to obtain (5.4.6) is a particular case of the following principle.

Principle of Optimality. An optimal policy has the property that whatever the initial state and initial decision are the remaining decisions must constitute an optimal policy with regard to the state resulting from the first decision.

If an optimal policy is conceived of as a means of tracing out an optimal trajectory in phase space, say a geodesic as far as time or distance are concerned, then the principle becomes quite clear:

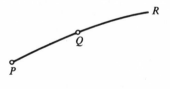

Figure 5.3

If PR is to be a path of minimum time, then clearly the last part, QR, must also be a path of minimum time.

Once Q has been determined, PQ must also be a geodesic. The point is, of course, that starting from P we don't always know the point Q in advance. This means that we can only assert that $PQ + QR$ must constitute an optimal trajectory.

Another way of putting this is to state that in a multistage decision process we must balance the gain or loss resulting from the first decision against the return from the optimal continuation from the remaining decisions.

In the remaining part of the chapter we will show how easy it is to apply the foregoing principle to treat a number of variational problems.

5.6. Discussion

So far, we do not seem to have gained much. Whereas the formalism of the calculus of variations yielded an ordinary differential equation, the new formalism confronts us with a partial differential equation of unfamiliar type. There is, however, one point to emphasize. The differential equation was subject to a two-point boundary condition, whereas the partial differential equation has its solution determined by an initial condition. These are very important points as far as analytic and computational solutions are concerned. We shall discuss some aspects of this in Chapter 8.

We can remove some of the strangeness by carrying out the minimization with respect to v. We have, differentiating in (5.4.9),

$$ v = -\frac{1}{2}\frac{\partial f}{\partial c}, \qquad (5.6.1) $$

and thus, using this value, we derive

$$ \frac{\partial f}{\partial T} = c^2 - \frac{1}{4}\left(\frac{\partial f}{\partial c}\right)^2, \qquad (5.6.2) $$

a nonlinear partial differential equation with the initial condition $f(c, 0) = 0$.

5.7. Simplification

We start with the observation that

$$f(c, T) = c^2 r(T). \tag{5.7.1}$$

We know this using the explicit form provided by the results of Chapter 4. As pointed out in that chapter, this representation is immediately apparent using the change of variable $u = cw$ and the quadratic nature of the integrand of $J(u)$. Using (5.7.1) in (5.6.2), we see that $r(T)$ satisfies the ordinary differential equation

$$r'(T) = 1 - r^2(T), \qquad r(0) = 0, \tag{5.7.2}$$

a Riccati equation.

EXERCISES

1. Solve (5.7.2) by means of separation of variables.
2. Solve it by conversion to a second-order linear differential equation with constant coefficients.

5.8. Validation

We have already established that

$$f(c, T) = c^2 \tanh T, \tag{5.8.1}$$

thus the function $r(T)$ of (5.7.2) is equal to $\tanh T$. We see that $r(T)$ does indeed satisfy (5.7.2). Alternatively, as indicated in the exercises, we can solve (5.7.2) explicitly and once again come up with the value $\tanh T$.

The point to emphasize is that we have once again invoked a familiar technique of analysis. A formal procedure based upon intuitive concepts is first used to obtain a relation. Once the form of the relation is observed, we can employ either an entirely differential method to establish it rigorously, or, guided by our knowledge of the desired result, return to the original method and smooth the edges.

After all, we have a choice of our techniques. No holy vow has been sworn to adhere rigidly to one or the other general theories. Successful mathematical research requires eclecticism.

5.9. Infinite Process

The formalism is particularly simple in the case where $T = \infty$. Write

$$f(c) = \min_{u} \int_0^\infty (u'^2 + u^2)\, dt. \qquad (5.9.1)$$

Then the same formalism as above yields the relation

$$f(c) = \min_{v} [(v^2 + c^2)\Delta + f(c + v\Delta)] + O(\Delta^2), \qquad (5.9.2)$$

whence

$$0 = \min_{v} [v^2 + c^2 + vf'(c)]. \qquad (5.9.3)$$

Hence,

$$v = -\frac{f'(c)}{2}, \qquad f'^2 = 4c^2. \qquad (5.9.4)$$

Since $f(c) \geq 0$ for $c \geq 0$, and is increasing as c increases, we see that $f'(c) = 2c$, whence

$$f(c) = c^2, \qquad v = -c. \qquad (5.9.5)$$

This is the same result we obtained previously as the limit of $f(c, T)$ as $T \to \infty$.

5.10. Limiting Behavior as $T \to \infty$

We discussed limiting behavior as $T \to \infty$ in Chapter 4 using the explicit form of $u(t) \equiv u(t, T)$. Let us now examine the question again, starting with (5.7.2).

If $r(T)$ approaches a limit as $T \to \infty$, we maintain that the limit is 1, a value we obtain by setting the right-hand side of (5.9.2) equal to zero.

We begin by showing that $r(T)$ is uniformly bounded for all $T \geq 0$. To establish this, consider the choice $u = ce^{-t}$, $u' = -ce^{-t}$ (which happens to be the solution for $T = \infty$; this is of no importance to the argument). Then

$$f(c, T) = \min_{u} J(u) \leq \int_0^T (c^2 e^{-2t} + c^2 e^{-2t})\, dt < 2c^2 \int_0^\infty e^{-2t}\, dt = c^2.$$

$$(5.10.1)$$

Thus, $f(c, T)$ is uniformly bounded by c^2 for $T \geq 0$, which means that $r(T) \leq 1$ for all $T \geq 0$. Next, we wish to show that $f(c, T)$ is monotone increasing as T increases. Let $u(t, T)$ denote the minimizing function. Then, if $T > T_1 \geq 0$,

$$f(c, T) = \int_0^{T_1} (u'(t, T)^2 + u(t, T)^2)\, dt + \int_{T_1}^T (\cdots)\, dt$$

$$\geq f(c, T_1) + \int_{T_1}^T (\cdots)\, dt > f(c, T_1).$$

$$(5.10.2)$$

We have thus established that $r(T)$ is uniformly bounded and monotone increasing in T. Consequently,

$$r(\infty) = \lim_{T \to \infty} r(T) \qquad (5.10.3)$$

exists. What is its value?

From the differential equation

$$r'(T) = 1 - r^2, \qquad r(0) = 0, \qquad (5.10.4)$$

we see that $r'(T)$ also possesses a limit as $T \to \infty$. This limit must be zero, otherwise

$$r(T) = \int_0^T r'(s)\, ds \qquad (5.10.5)$$

would be unbounded as $T \to \infty$. If $r'(\infty) = 0$, we see from the differential equation that $1 - r^2(\infty) = 0$. Hence, $r(\infty) = \pm 1$. Since $r(\infty) > 0$, the correct limiting value is 1.

EXERCISE

1. Show that $r(T) = 1 + O(e^{-2T})$ as $T \to \infty$, without using the explicit value of $r(T)$. (*Hint:* Write $r = 1 + s$ in (5.10.4) and consider the equation for s.)

5.11. Two-Point Boundary Problems

Let us now consider the problem of minimizing the functional $J(u) = \int_0^T (u'^2 + u^2)\, dt$ subject to the constraints $u(0) = c$, $u(T) = c_2$. Writing once again

$$f(c, T) = \min_u J(u), \tag{5.11.1}$$

there is no difficulty in verifying that

$$f_T = \min_v \left[v^2 + c^2 + v f_c \right]$$

$$= c^2 - \frac{f_c^2}{4}. \tag{5.11.2}$$

What, however, is the initial condition as $T \to 0$?

The point of difficulty is that $f(c, T)$ is not continuous in a closed interval $[0, T_0]$ as a function of T, and indeed is not bounded as $T \to 0$. For T small, we know that

$$u \cong c + \frac{(c_2 - c)t}{T}, \qquad u' \cong \frac{c_2 - c}{T} \tag{5.11.3}$$

(the straight line approximation). Hence, for T small,

$$f(c, T) \cong \int_0^T \left[\frac{(c_2 - c)^2}{T^2} + \left(c_1 + \frac{(c_2 - c)t}{T} \right)^2 \right] dt$$

$$\cong \frac{(c_2 - c)^2}{T}. \tag{5.11.4}$$

This is the initial condition to be used in solving (5.11.2). As the exercises will show, we can go further in the asymptotic characterization of the solution as $T \to 0$.

EXERCISES

1. Use the relation $f(c, T) = c^2 r_1(T) + 2c_2 cr_2(T) + c_2^2 r_3(T)$ to obtain a system of ordinary differential equations for r_1, r_2, r_3.
2. Use (5.11.4) to obtain the suitable initial conditions and thus determine r_1, r_2, r_3. Verify the answers using the results of Chapter 4.

5.12. Time-Dependent Control Process

Let us next consider the problem of minimizing the functional

$$J(u) = \int_0^T [u'^2 + g(t)u^2] \, dt. \qquad (5.12.1)$$

Because of the time-dependence of the integrand, it is no longer true that a control process over $[\Delta, T]$ is equivalent to the same control process over $[0, T - \Delta]$. Consequently, the introduction of the function $f(c, T)$ as above is of no particular use.

Instead, we introduce our imbedding by means of the new function

$$\varphi(c, a) = \min_u \int_a^T [u'^2 + g(t)u^2] \, dt, \qquad (5.12.2)$$

where $u(a) = c$ and $0 \leq a \leq T$. We run time backwards instead of forwards. Then, arguing as before, we readily obtain the equation

$$-\varphi_a = \min_v [v^2 + g(a)c^2 + v\varphi_c], \qquad (5.12.3)$$

whence

$$-\varphi_a = g(a)c^2 - \varphi_c^2/4, \qquad \varphi(c, T) = 0. \qquad (5.12.4)$$

Similarly, we set $\varphi(c, a) = c^2 \rho(a)$ and obtain the Riccati equation

$$-\rho'(a) = g(a) - \rho^2(a), \qquad \rho(T) = 0. \qquad (5.12.5)$$

Although we can no longer expect an explicit analytic solution, in the general case where $g(a)$ is not constant, we have achieved the end of replacing a two-point boundary value problem by an initial value problem.† Furthermore, we obtain the return function and the optimal policy directly from this equation.

† We are thinking ahead to the use of a digital computer, as discussed in more detail in Chapter 8.

EXERCISES

1. Use the results of Section 20 of Chapter 4 to verify that $\rho(a)$ satisfies (5.12.5).
2. Given the function $\rho(a)$, what is the equation of the extremal function, that is, the function that minimizes the quadratic functional.
3. What is the initial condition at $a = T$ if no constraint is imposed at $a = T$? What if there is a condition $u(T) = c_2$?

5.13. Global Constraints

Let us now consider the problem of maximizing $J_1(u) = \int_0^T u^2 \, dt$ subject to the constraint $\int_0^T u'^2 \, dt = k$, $u(0) = c$. As before, we can form the functional

$$J(u) = \int_0^T (u^2 - \lambda u'^2) \, dt, \qquad (5.13.1)$$

where $\lambda > 0$, and then apply the dynamic programming formalism.

There is, however, another approach that is more direct. Let us consider k to be another state variable.† A choice of $u' = v$ at $t = 0$ then has three effects,

$$\begin{aligned} c &\rightarrow c + v\Delta, \\ k &\rightarrow k - \Delta v^2, \\ T &\rightarrow T - \Delta. \end{aligned} \qquad (5.13.2)$$

where Δ again is a small quantity and we ignore terms of order Δ^2. If we write

$$f(c, k, T) = \max J_1(u), \qquad (5.13.3)$$

we have, employing the principle of optimality,

$$f(c, k, T) = \max_v \, [c^2\Delta + f(c + v\Delta, k - v^2\Delta, T - \Delta)] + O(\Delta^2), \qquad (5.13.4)$$

---

† In many applications it corresponds to a limited quantity of some resource, for example, energy.

whence, passing to the limit as Δ approaches zero, we obtain the partial differential equation

$$f_T = \max_v \left[c^2 + vf_c - v^2 f_k \right]. \tag{5.13.5}$$

The initial condition with respect to k is simple. If $k = 0$, then it follows that $u'(t) = 0$, $0 \le t \le T$ and $J_1(u) = c^2 T$. On the other hand, if T approaches zero, then $u' \sim (k/T)^{1/2}$, $u \sim c + t(k/T)^{1/2}$, for $0 \le t \le T$, whence

$$u^2 \sim c^2 + 2ct \left(\frac{k}{T} \right)^{1/2} + \frac{t^2 k}{T}. \tag{5.13.6}$$

Thus, $J_1(u) \sim c^2 T$ as $T \to 0$, which shows that $f(c, k, T)$ is continuous as k and T approach zero independently.

We have proceeded formally in what preceded since we are not interested in applying any of these results. What we do wish to point out is that if we compare (5.13.5) and the equation obtained from (5.13.1), namely,

$$f_T = \max_v \left[c^2 + vf_c - \lambda v^2 \right], \tag{5.13.7}$$

then we see the correspondence between λ and f_k. The quantity f_k is a "price" in the usual sense of the term in economics. That the Lagrange multiplier may be considered to be a price is a standard result in mathematical economics. We shall discuss local constraints such as $|u'| \le k$ in a later section.

EXERCISES

1. Write $f(c, k, T) = c^2 T + f_1(c, T)k^{1/2} + f_2(c, T)k + \cdots$, and obtain the partial differential equations for the coefficients, f_1, f_2, and so on.
2. Make the change of variable $u = cw$ and deduce that $f(c, k, T) = c^2 f(1, k/c^2, T)$. Make the change of variable $u = k^{1/2} z$ and deduce that $f(c, k, T) = kf(c/k^{1/2}, 1, T)$.
3. Hence reduce (5.13.5) to a partial differential equation involving only two variables.
4. Solve (5.13.5) for f_k.
5. What is the expansion for $f(c, k, T)$ in terms of c?

6. Consider $\lambda_1 = \min \int_0^T u'^2\, dt$ over $u(0) = u(T) = 0$, $\int_0^T u^2\, dt = 1$. Imbed this determination of the smallest characteristic value of the equation $u'' + \lambda u = 0$, $u(0) = u(T) = 0$ in the general problem of minimizing $\int_0^T u'^2\, dt$ subject to $u(0) = c$, $u(T) = 0$, $\int_0^T u^2\, dt = 1$, and obtain a partial differential equation satisfied by

$$f(c, T) = \min \int_0^T u'^2\, dt.$$

What is the behavior of $f(c, T)$ as $T \to 0$?

7. Consider the corresponding problem for the equation $u'' + \lambda \varphi(t) u = 0$, $u(0) = u(T) = 0$.

5.14. Discrete Control Processes

In many situations, we cannot control a system in a continuous fashion. It may not be possible to observe it continuously, nor to make an unbroken succession of decisions. Furthermore, it is of considerable interest to determine whether such close scrutiny of a system is necessary. It may be the case that intermittent control will be just as effective, and much cheaper and simpler to implement. Esthetically, there is a great advantage to formulating a mathematical model directly in discrete terms if we expect to use a digital computer to obtain a numerical solution of the resultant equations. From the mathematical point of view, there is the advantage that a direct rigorous approach can be made without any worry about the subtleties of the calculus of variations.

Let us then consider a system S whose state at time n, $n = 0, 1, 2, \ldots$, is described by the scalar variable u_n. Let the equation determining the history of the system be

$$u_{n+1} = au_n + v_n, \qquad u_0 = c, \tag{5.14.1}$$

and suppose that the v_n, the control variables, are chosen so that the quadratic form

$$J_N(u, v) = \sum_{n=0}^{N} (u_n^2 + v_n^2) \tag{5.14.2}$$

is minimized. This is the discrete analogue of the first problem discussed in Chapter 4.

5.15. Preliminaries

Since the minimum value of $J(u, v)$ depends on c, the initial value, and N, the number of stages, we write

$$f_N(c) = \min_{\{v_n\}} J_N(u, v). \tag{5.15.1}$$

Does the minimum actually exist or should we write "inf"? Proceeding as in Chapter 4, we can readily answer that an absolute minimum is attained. However, we can also argue from first principles. Suppose that we take $v_n = 0$, $n = 0, 1, 2, \ldots, N$. The corresponding value of $J_N(u, 0)$ is $\sum_{n=0}^{N} a^2 c^{2n}$. Hence, we know that

$$\min_{\{v_n\}} J_N(u, v) \leq c^2 \sum_{n=0}^{N} a^{2n}. \tag{5.15.2}$$

Thus, since

$$\sum_{n=0}^{N} v_n^{2} < \sum_{n=0}^{N} (u_n^{2} + v_n^{2}),$$

it is sufficient to ask for the minimum of $J_N(u, v)$ over the finite region

$$\sum_{n=0}^{N} v_n^{2} \leq c^2 \sum_{n=0}^{N} a^{2n+2}. \tag{5.15.3}$$

The minimum of $J_N(u, v)$ is assumed over this closed region and thus $f_N(c)$ exists as a function of c for all c and $N = 0, 1, 2, \ldots$. As before, the change of variable $v_n = cz_n$, $u_n = cw_n$, shows directly that $f_N(c) = c^2 r_N$, where r_N is independent of c.

5.16. Recurrence Relation

Let us now apply the principle of optimality to obtain a recurrence relation, a nonlinear difference equation, satisfied by $f_N(c)$.

After v_0 is chosen, the new state of the system is $ac + v_0$. The criterion function takes the form

$$c^2 + v_0^{2} + \sum_{n=1}^{N} (u_n^{2} + v_n^{2}). \tag{5.16.1}$$

The additivity of the integrand shows that regardless of the choice of v_0, we proceed from the new state to minimize the remaining sum. Hence, we know that for any choice of v_0, the criterion function takes the form

$$c^2 + v_0{}^2 + f_{N-1}(ac + v_0). \qquad (5.16.2)$$

The quantity v_0 is now to be chosen to minimize the expression. Thus, we obtain the nonlinear recurrence relation

$$f_N(c) = \min_{v_0} [c^2 + v_0{}^2 + f_{N-1}(ac + v_0)], \qquad (5.16.3)$$

$N \geq 1, f_0(c) = c^2$.

It can now be established inductively that $f_N(c) = c^2 r_N$, where r_N is independent of c. We have already established this by a change of variable in the previous section. Arguments completely analogous to the continuous case show that r_N is monotone increasing in N and uniformly bounded. Hence, r_N converges as $N \to \infty$.

5.17. Explicit Recurrence Relations

Using the fact that $f_N(c) = c^2 r_N$, (5.16.3) yields the relation

$$c^2 r_N = \min_v [(c^2 + v^2) + r_{N+1}(ac + v)^2]. \qquad (5.17.1)$$

The value of $v, \bar{v}$ that minimizes is readily obtained by differentiation,

$$2\bar{v} + 2r_{N-1}(ac + \bar{v}) = 0,$$

$$\bar{v} = \frac{-r_{N-1}ac}{1 + r_{N-1}}. \qquad (5.17.2)$$

Using this value and a small amount of algebra in (5.17.1), we obtain the relation

$$r_N = (1 + a^2) - \frac{a^2}{(1 + r_{N-1})} \qquad (5.17.3)$$

for $N \geq 1$ with $r_0 = 1$. This is a discrete analogue of the Riccati equation obtained in the continuous case.

5.18. Behavior of r_N

It is clear from (5.17.3) that r_N is uniformly bounded by the quantity $(1 + a^2)$, and that it is monotone increasing. We have

$$r_1 = 1 + a^2 - \frac{a^2}{2} > 1 = r_0 \, ,$$

$$r_2 = 1 + a^2 - \frac{a^2}{1 + r_1} > 1 + a^2 - \frac{a^2}{1 + r_0} = r_1 \, ,$$

(5.18.1)

and thus, inductively, $r_N > r_{N-1}$. This monotone behavior in N follows, of course, directly from the definition of $f_N(c)$.

Let $r = \lim_{N \to \infty} r_N$. Then r is the positive root of the quadratic equation

$$r = 1 + a^2 - \frac{a^2}{1 + r} \, .$$

(5.18.2)

Observe that $\bar{v}$, the initial decision, also converges as $N \to \infty$, and

$$\lim_{N \to \infty} \bar{v} = - \frac{rac}{1 + r} \, .$$

(5.18.3)

If we formally consider the infinite process,

$$f(c) = \min_{\{v_n\}} \sum_{n=0}^{\infty} (u_n^2 + v_n^2),$$

(5.18.4)

we see that

$$f(c) = \min_{v} [c^2 + v^2 + f(ac + v)].$$

(5.18.5)

One solution of this equation is certainly $f(c) = rc^2$. Since we are not particularly interested in the infinite process, we shall table any discussion of appropriate conditions to impose upon $f(c)$ to ensure uniqueness.

EXERCISES

1. Is the solution of (5.18.5) unique under the condition that $f(c)$ is a quadratic in c?

2. Is the solution unique under the condition that $f(c)$ is strictly convex?

3. Is it unique if we suppose that $f(c)$ is differentiable for $c \geq 0$ and that $0 \leq f(c) \leq kc^2$?

5.19. Approach to Steady-State Behavior

Let us now investigate the rapidity with which the steady-state control control policy is assumed. Write

$$r_N = r - w_N. \tag{5.19.1}$$

Then

$$r - w_N = 1 + a^2 - \frac{a^2}{1 + r - w_{N-1}}$$

$$= 1 + a^2 - \frac{a^2}{1 + r} - \frac{a^2}{(1 + r)^2} w_{N-1} + O(w_{N-1}^2). \tag{5.19.2}$$

Hence, asymptotically as $N \to \infty$,

$$w_N \cong \frac{a^2}{(1 + r)^2} w_{N-1}. \tag{5.19.3}$$

Since $r_N \to r$ as $N \to \infty$, we must have $a^2 < (1 + r)^2$ and this is easily verified. Thus, w_N approaches zero exponentially as $N \to \infty$. In the next section, we will demonstrate this by means of an explicit calculation of the solution of (5.19.2).

5.20. Equivalent Linear Relations

In the relation

$$r_N = 1 + a^2 - \frac{a^2}{1 + r_{N-1}}, \tag{5.20.1}$$

set $r_N = p_N/q_N$. Then (5.20.1) becomes

$$\frac{p_N}{q_N} = \frac{q_{N-1} + (1 + a^2)p_{N-1}}{q_{N-1} + p_{N-1}}. \tag{5.20.2}$$

Hence, if we set

$$p_N = (1 + a^2)p_{N-1} + q_{N-1}, \qquad p_0 = 1,$$
$$q_N = p_{N-1} + q_{N-1}, \qquad\qquad q_0 = 1, \qquad (5.20.3)$$

we know that r_N, as given by p_N/q_N, satisfies (5.20.1). Since (5.20.3) is a linear system, the analytic solution can readily be obtained.

EXERCISES

1. Find the explicit solution of (5.20.3) and thus determine the asymptotic behavior of r_N.
2. Show that r_N can also be obtained from the solution of $p_N = k[(1 + a^2)p_{N-1} + q_{N-1}]$, $q_N = k[p_{N-1} + q_{N-1}]$, $p_0 = q_0 = 1$ for any $k \neq 0$. Is there any advantage to choosing k different from one?

5.21. Local Constraints

The problem of determining optimal control subject to constraints is quite difficult, as mentioned in the previous chapter. Let us, without dwelling upon the rigorous aspects, indicate briefly how we modify the preceding formalism to treat some more general optimization problems. Consider, for example, the question of minimizing the functional

$$J(u) = \int_0^T (u'^2 + u^2)\, dt \qquad (5.21.1)$$

subject to the initial condition and the constraint

$$|u'| \leq 1. \qquad (5.21.2)$$

Writing $f(c, T) = \min_u J(u)$, we obtain in the usual formal manner the nonlinear partial differential equation

$$f_T = \min_{|v| \leq 1} [c^2 + v^2 + vf_c], \qquad f(c, 0) = 0, \qquad (5.21.3)$$

an equation that is considerably more difficult to treat. We shall discuss problems of this type in a subsequent volume.

In passing, let us mention that it is not at all obvious that the function $f(c, T)$ possesses the required partial derivatives, and, as a matter of fact, (5.21.3) has to be interpreted properly. There are several ways of circumventing these difficulties, one of which we shall discuss below.

If the constraint reads $|u'| \le k|u|$, we obtain

$$f_T = \min_{|v| \le k|c|} [c^2 + v^2 + vf_c]. \tag{5.21.4}$$

Setting $u = cw$, we see that $f = c^2 r(T)$, where $r(T)$ is determined by the nonlinear equation

$$r'(T) = \min_{|w| \le k} [1 + w^2 + 2wr(T)], \qquad r(0) = 0. \tag{5.21.5}$$

Assuming for the moment that (5.21.3) is a valid equation, let us show that it can readily be used to determine the solution. Consider the minimum of the expression $c^2 + v^2 + vf_c$ with respect to v. If $|f_c| < 2$, the minimum is assumed at $v = -f_c/2$. If $|f_c| \ge 2$, the minimum is assumed for $v = \pm 1$ ($+1$ if $f_c < -2$, -1 if $f_c > 2$).

We suspect that f is monotone increasing as $|c|$ increases and as T increases. Hence, for c and T small, the minimum will be assumed at $v = -f_c/2$ and $f(c, T)$ will have the same analytic form as before. Thus, we suspect that there is a region in the (c, T)-plane within which $f(c, T) = c^2 r(T)$. The boundary to this region should be determined by the condition

$$cr(T) = 1, \tag{5.21.6}$$

considering only the case $c > 0$. Since $r(\infty) = 1$, the line $c = 1$ is an asymptote to the curve described by (5.21.6).

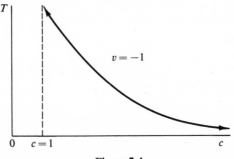

Figure 5.4

One of the advantages of a formal procedure of the type sketched above is that a solution once guessed can often be readily validated.

5.22. Continuous as Limit of Discrete

One way of avoiding difficulties introduced by constraints as far as the existence of solutions with the desired partial derivatives are concerned is to consider an associated discrete process. Let u_n be the state variable and suppose that

$$u_{n+1} = u_n + (au_n + v_n)\Delta, \qquad u_0 = c, \qquad (5.22.1)$$

and that it is desired to minimize

$$J_N = \sum_{n=0}^{N} (u_n^2 + v_n^2)\Delta \qquad (5.22.2)$$

subject to the constraint

$$|v_n| \le 1, \qquad n = 0, 1, \dots. \qquad (5.22.3)$$

Writing

$$f_N(c) = \min_{\{v_n\}} J_N, \qquad (5.22.4)$$

we readily obtain the functional equation

$$f_N(c) = \min_{|v| \le 1} [(c^2 + v^2)\Delta + f_{N-1}(c + (ac + v)\Delta)], \qquad (5.22.5)$$

for $N \ge 1$, with $f_0(c) = 0$.

The analytic structure of the solution of this equation can readily be obtained by means of an inductive argument. Two questions now arise. If we write $f_N(c) = f_N(c, \Delta)$ and let Δ approach zero, through some sequence of values, is it true that $\{f_N(c, \Delta)\}$ converges? Secondly, if it converges, does it converge to the solution of the corresponding continuous variational problem?

These are important problems from both the analytic and computational point of view, which we shall discuss in some detail in Volume II.

5.23. Bang-Bang Control

Suppose that we do our best to thwart calculus by requiring that u' assume only the values ± 1. We think of u' having the value $+1$ over $[0, t_1]$, -1 over $[t_1, t_2]$, and so on, or alternatively starting out with the value -1. It is clear that quite pathological control policies can be generated in this fashion if we allow an infinite number of alternations.

Assume that we allow at most N changes in control policy. Let

$$f_N(c, +, T) = \min \int_0^T (u'^2 + u^2) \, dt, \qquad (5.23.1)$$

where $u(0) = c$, u' starts with the values $+1$ and at most N switchings are allowed over $[0, T]$. Similarly, we define $f_N(c, -, T)$.

Then it is easy to compute $f_0(c, +, T), f_0(c, -, T)$. Namely,

$$f_0(c, +, T) = \int_0^T (1 + (c + t)^2) \, dt,$$

$$\qquad (5.23.2)$$

$$f_0(c, -, T) = \int_0^T (1 + (c - t)^2) \, dt.$$

If the first switch is at t_1, we have

$$f_N(c, +, T) = \min_{0 \le t_1 \le T} \left[\int_0^{t_1} (1 + (c + t)^2) \, dt + f_{N-1}(c + t_1, -, T - t_1) \right],$$

$$f_N(c, -, T) = \min_{0 \le t_1 \le T} \left[\int_0^{t_1} (1 + (c - t)^2) \, dt + f_{N-1}(c - t_1, +, T - t_1) \right],$$

$$\qquad (5.23.3)$$

for $N = 1, 2, \ldots$. We then use the minimum of $f_N(c, +, T)$ and $f_N(c, -, T)$ to determine the desired minimum.

If there is a cost for each switching, say s, we can determine the most efficient procedure by minimizing the functions $f_N(c, \pm, T) + sN$ over N.

1. Suppose that the state variable must have the form

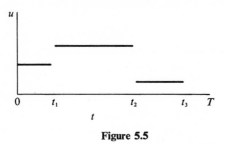

Figure 5.5

where N switch points $t_1, t_2, \ldots, t_N$ are allowed and the values of the levels are arbitrary after the first, and that the cost function is

$$\int_0^T u^2 \, dt + \sum_{i=0}^{N-1} g(t_{i+1} - t_i).$$

Obtain a functional equation for the minimum value and solve explicitly if $g(s) = s^2$.

5.24. Control in the Presence of Unknown Influences

Let us now once again consider the problem of controlling a system in the presence of unknown disturbing influences. Consider first the case of discrete control. Let

$$u_{n+1} = au_n + v_n + w_n, \qquad u_0 = c. \tag{5.24.1}$$

where the v_n are chosen to minimize and the w_n to maximize the quadratic form

$$Q_N = \sum_{n=0}^{N} (u_n{}^2 + v_n{}^2 - w_n{}^2). \tag{5.24.2}$$

Consider first the situation where the v_n and w_n are chosen in the following fashion. First v_0, then w_0; then v_1, then w_1, and so on.

If we write

$$f_N(c) = \min_{v_0} \max_{w_0} \left[\min_{v_1} \max_{w_1} [\cdots] \right], \qquad (5.24.3)$$

we see that for $N \geq 1$,

$$f_N(c) = \min_v \max_w [c^2 + v^2 - w^2 + f_{N-1}(ac + v + w)], \qquad (5.24.4)$$

with $f_0(c) = c^2$.

We leave it to the reader to show that $f_N(c) = r_N c^2$ and to obtain a recurrence relation for r_N.

On the other hand, the process could proceed in the following fashion: first w_0, then v_0; then w_1, then v_1, and so on.

If we write

$$F_N(c) = \max_{w_0} \min_{v_0} \left[\max_{w_1} \min_{v_1} [\cdots] \right], \qquad (5.24.5)$$

we see that for $N \geq 1$,

$$F_N(c) = \max_w \min_v [c^2 + v^2 - w^2 + F_{N-1}(ac + v + w)], \qquad (5.24.6)$$

with $F_0(c) = c^2$.

We leave it to the reader to demonstrate that $F_N(c) = f_N(c)$.

Generally, a direct calculation will show that it makes no difference in this case in what order the minima and maxima with respect to the v_n and w_n are taken.

5.25. Comparison between Calculus of Variations and Dynamic Programming

In this and the preceding chapter, we employed two methods apparently quite different in conceptual and analytic structure. Typical of the inner unity that exists in mathematics is the fact that these two theories, the calculus of variations and dynamic programming, are actually intimately related; they are dual theories.

It would take us too far out of our way to discuss in detail what we mean by this. Let us merely indicate one facet of this duality.

In the calculus of variations, we wish to minimize a functional $J(u)$. The minimizing function is regarded as a point in a function space; see Fig. 5.6. In the dynamic programming approach, the tangent at a point is determined by the policy; see Fig. 5.7.

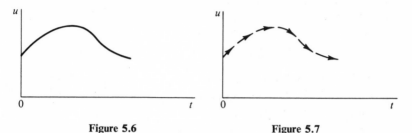

Figure 5.6   **Figure 5.7**

Thus, in the calculus of variations we regard a curve as a locus of points, while in dynamic programming we regard it as an envelope of tangents. The basic duality theorem of Euclidean space is that these are equivalent formulations.

Just as in projective geometry, the choice of which formulation to employ at a particular time is a matter of convenience. These matters will be explored in depth in subsequent volumes. One additional point should be mentioned here. This correspondence disappears when we turn to stochastic control processes. There the two approaches correspond to quite different physical situations with information the key.

Miscellaneous Exercises

1. Consider the problem of minimizing $J(u) = \int_0^T (u'^2 + u^2)\, dt$, where $u(0) = c$. Show directly, without knowledge of the explicit solution, that $|u| \le |c| g(T)$ for an appropriate function $g(T)$.

2. Consider the quadratic functional

$$J(u, v) = \int_a^1 [q(t)u^2 - p(t)u'^2 - 2uv(t)]\, dt - \frac{p(1)u(1)^2}{\alpha}.$$

Show that the minimum over all u satisfying $u(a) = c$, $u(1) + \alpha u'(1) = 0$ exists provided that the smallest characteristic value of the Sturm-Liouville equation $(pu)' + \lambda qu = 0$, $u(a) = 0$, $u(1) + \alpha u'(1) = 0$ is larger than one. Show that this is the case if $d \geq q \geq 0$ with d sufficiently small, or if $1 - a$ is sufficiently small.

3. Show that the associated Euler equation is $(pu')' + qu = v$, $u(a) = c$, $u(1) + \alpha u'(1) = 0$.

4. Write $f(a, c) = \max_a J(u, v)$. Show that

$$-\frac{\partial f}{\partial a} = \left(\frac{\partial f}{\partial c}\right)^2 \bigg/ 4p(a) - 2cv(a) + c^2 q(a).$$

What is the limiting behavior of $f(a, c)$, as a approaches 1?

5. Show that the Green's function associated with the equation in Exercise 3 is given by

$$K(s, t, a) = \frac{1}{p(a)} \left[\frac{G(u_1)u_2(t) - u_1(t)G(u_2)}{G(u_2)}\right] u_2(s), \qquad a < t < s,$$

$$= \frac{1}{p(a)} \left[\frac{G(u_1)u_2(s) - u_1(s)G(u_2)}{G(u_2)}\right] u_2(t), \qquad a < s < t,$$

where u_1, u_2 are the principal solutions and $G(u) = u(1) + \alpha u'(1)$.

6. Let $\varphi(t)$ be the solution of $(pu') + qu = 0$, $u(a) = 1$, $u(1) + \alpha u'(1) = 0$. Show that $\varphi(t) = -p(a)[(\partial K/\partial s)(s, t, a)]_{s=a}$. Henceforth, we shall write this last expression as $(\partial K/\partial s)(a, t, a)$.

7. Show that

$$f(a, c) = -\int_a^1 \int_a^1 K(s, t, a)v(s)v(t)\, ds\, dt$$

$$- 2c \int_a^1 \varphi v\, dt + c^2 p(a)\varphi'(a).$$

8. Let $F(s, t)$ be continuous in $a \leq s \leq 1$, $a \leq t \leq 1$, and symmetric, $F(s, t) = F(t, s)$. If

$$\int_a^1 \int_a^1 F(s, t)v(s)v(t)\, ds\, dt = 0$$

for all continuous functions v, then $F(s, t) \equiv 0$.

9. Using the partial differential equation of Exercise 3, show that

$$\frac{\partial K}{\partial a}(s, t, a) = \frac{\varphi(s)\varphi(t)}{p(a)} = p(a)\frac{\partial K}{\partial a}(s, a, a)\frac{\partial K}{\partial a}(a, t, a).$$

10. Consider the equation $(pu')' + (z + q(t))u = 0$, where z is a complex variable. Show that the principal solutions are entire functions of z for any value of t and that $K(s, t, a, z)$ is a meromorphic function of z for any value of s and t.

11. Show that the maximum of

$$J(u, v, z) = \int_a^1 [(z + q(t))u^2 - p(t)u'^2 - 2uv(t)] \, dt - \frac{p(1)u(1)^2}{\alpha}$$

exists if z is a large enough negative constant. Hence, show that the relations of the foregoing exercises hold if $p(t) > 0$, $q(t)$ is continuous, and z is a large negative quantity. Hence, show that the foregoing relations hold whenever both sides are finite, using the meromorphic character of φ and K as functions of z.

12. Consider the equation $(pu')' + (q + \lambda r)u = v$, $a < t < 1$, $u(a) = 0$, $u(1) + \alpha u'(1) = 0$. If λ is not a characteristic value, this has a unique solution $u(t) = \int_a^1 R(t, s, \lambda, a)v(s) \, ds$. This function R is called the resolvent. Let $\psi_k(t)$ be the characteristic function associated with the characteristic value λ_k. Show that

$$R(t, s, \lambda, a) = \sum_{k=1}^\infty \frac{\psi_k(t)\psi_k(s)}{\lambda - \lambda_k}.$$

13. Using this last result, plus that of Exercise 9, show that

$$\frac{\partial \lambda_k}{\partial a} = p(a)\psi_k'(a)^2,$$

$$\frac{\partial \psi_k(t)}{\partial a} = p(a)\psi_k'(a) \sum_{j \neq k} \frac{\psi_j(t)\psi_j'(a)}{\lambda - \lambda_j}.$$

(For the preceding, see

R. Bellman and S. Lehman, "Functional Equations in the Theory of Dynamic Programming—X: Resolvents, Characteristic Functions, and Values," *Duke Math. J.*, **27**, 1960, pp. 55–69.
K. S. Miller and M. M. Schiffer, "On the Green's Functions of Ordinary Differential Systems," *Proc. Amer. Math. Soc.*, **3**, 1952, pp. 433–441.

For an extension to differential equations involving complex coefficients involving min-max operations, see

R. Bellman and S. Lehman, "Functional Equations in the Theory of Dynamic Programming—XII: Complex Operators and Min-Max Variation," *Duke Math. J.*, **28**, 1961, pp. 335–343.)

14. Consider the equation $u' = g(u)$, $u(0) = c$, where $g(u)$ is convex. Then It may be written $u' = \max_v [g(u) + (u - v)g'(v)]$, $u(0) = c$, where a choice of v corresponds to a policy. Show that the Newton-Raphson-Kantorovich approximation scheme

$$u'_{n+1} = g(u_n) + (u_{n+1} - u_n)g'(u_n), \qquad u_{n+1}(0) = c,$$

$n = 0, 1, 2, \ldots$, with u_0 specified, corresponds to an approximation in policy space in which the nth policy is taken to be the $(n - 1)$st return function.

15. Show that the approximation is monotone, $u_1 \leq u_2 \leq \cdots \leq u_n \leq u$, within the interval of definition of the solution $u(t)$.

16. Show that the convergence is quadratic, that is,

$$\max_{0 \leq t \leq t_0} |u - u_n| \leq k_1 \max_{0 \leq t \leq t_0} |u - u_{n-1}|^2.$$

17. Is there any advantage in using a different approximation in policy space for small n; for example, $v_n = (u_1 + u_2 + \cdots + u_{n-1})/(n - 1)$?

18. Use the fact that if $u'' - a(t)u = 0$, then $v = u'/u$ satisfies the Riccati equation $v' + v^2 - a(t) = 0$ to obtain a sequence of monotone approximations to u. If $a(t) > 0$, consider the equation for v^{-1} to obtain a sequence of monotone approximations to v^{-1} and thus to u from the other direction.

(For applications of the foregoing idea to quantum mechanical scattering, see

F. Calogero, *Variable Phase Approach to Potential Scattering*, Academic Press, New York, 1967.

See also

R. Bellman, "Functional Equations in the Theory of Dynamic Programming—V: Positivity and Quasilinearity," *Proc. Nat. Acad. Sci. U.S.A.*, **41**, 1955, pp. 743–746.

R. Bellman and R. Kalaba, *Quasilinearization and Nonlinear Boundary-Value Problems*, American Elsevier, New York, 1965.)

19. Consider the problem of minimizing the quadratic expression

$$Q(x) = \sum_{k=1}^{N} c_k(x_k - x_{k-1})^2 + \sum_{k=1}^{N} d_k(x_k - a_k)^2,$$

over the quantities $x_1, x_2, \ldots, x_N$, where x_0 is a given constant and the c_k and d_k are nonnegative. Consider the general problem of minimizing the quadratic form

$$Q_r(x) = \sum_{k=r}^{N} c_k(x_k - x_{k-1})^2 + \sum_{k=r}^{N} d_k(x_k - a_k)^2,$$

$r = 1, 2, \ldots, N$. Write $f_r(u) = \min Q_r(x)$, where $x_{r-1} = u$ and the minimum is taken over the remaining x_i. Show that

$$f_N(u) = \min_{x_N} [c_N(x_N - u)^2 + d_N(x_N - a_N)^2],$$

$$f_r(u) = \min_{x_r} [c_r(x_r - u)^2 + d_r(x_r - a_r)^2 + f_{r+1}(x_r)],$$

for $r = 0, 1, \ldots, N-1$.

20. Show that $f_r(u) = u_r + v_r u + w_r u^2$, where u_r, v_r, w_r depend only on r, and use the foregoing recurrence relation to obtain recurrence relations connecting u_r, v_r, w_r with u_{r+1}, v_{r+1}, w_{r+1}.

21. Treat in a similar fashion the problem of minimizing

$$Q(x) = \sum_{k=1}^{N} c_k(x_k - x_{k-1})^2 + \sum_{k=1}^{N} d_k(x_k - a_k)^2$$

$$+ \sum_{k=1}^{N} e_k(x_k - 2x_{k-1} + x_{k-2})^2,$$

where $x_0 = u$, $x_{-1} = v$.

22. How do we solve the problem of minimizing over all of the x_i, $i = -1, 0, 1, \ldots, N$?

(For the foregoing, see

R. Bellman, *Introduction to Matrix Analysis*, McGraw-Hill, New York, 1960.)

23. Consider the problem of minimizing the quadratic functional $J(u) = \int_0^T (u'^2 + \varphi(t)u^2)\,dt$, and let us employ the following approximation in policy space:

$$u'(t) = v_n, \qquad n\Delta \le t \le (n+1)\Delta, \qquad n = 0, 1, \ldots, (N-1)\Delta,$$

where the v_n are to be chosen to minimize, and suppose that $u(t)$ is continuous at the transition points $t = n\Delta$. Let $u_n = u(n\Delta)$, $n = 0, 1, \ldots, (N-1)\Delta$. Here $u_{n+1} - u_n = v_n \Delta$. Write

$$J(u, \Delta) = \sum_{n=0}^{N-1} \int_{n\Delta}^{(n+1)\Delta} (u_n{}^2 + \varphi(t)(u_n + (t - n\Delta)v_n)^2)\,dt.$$

Show that $\min_{\{v_n\}} J(u, \Delta) \ge \min_{\{v_n\}} J(u, \Delta/2) \ge \min_u J(u)$.

24. Use the recurrence relations of the types appearing above to obtain an expression for $\min_{\{v_n\}} J(u, \Delta)$.

25. For the case where $\varphi(t) = 1$, show by means of direct calculation that $\lim_{\Delta \to 0} \min_{\{v_n\}} J(u, \Delta) = \min_u J(u)$.

26. Carry through the proof in the general case under the assumption that $\varphi(t)$ is continuous and nonnegative for $0 \le t \le T$.

27. Under the same assumptions, consider the associated problems of minimizing

$$J_1 = \sum_{n=1}^{N} (u_n - u_{n-1})^2 \Delta + \sum_{n=1}^{N} u_{n-1}^2 \varphi((n-1)\Delta)\Delta,$$

and

$$J_2 = \sum_{n=1}^{N} (u_n - u_{n-1})^2 \Delta + \sum_{n=1}^{N} u_{n-1}^2 \varphi(n\Delta)\Delta.$$

28. Let

$$f_r(v) = \min_{\{u_k\}} \sum_{k=r}^{N} (u_k - u_{k-1})^2,$$

where $u_{r-1} = v$, and the u_k are subject to the conditions $u_N = 0$, $\sum_{k=r}^{N-1} \varphi_k u_k^2 = 1$, with $\varphi_k > 0$. Show that

$$f_{N-1}(v) = (\varphi_{N-1}^{-1/2} - v)^2,$$

$$f_r(v) = \min_{u_r} \left[(u_r - v)^2 + (1 - \varphi_r u_r^2) f_{r+1}\left(\left(\frac{v}{1 - \varphi_r u_r}\right)^2\right) \right],$$

for $r = 1, 2, \ldots, N-1$.

29. Treat in the same fashion the problem of determining the maximum and minimum of

(a) $(ax_1)^2 + (x_1 + ax_2)^2 + \cdots + (x_1 + x_2 + \cdots + x_{N-1} + ax_N)^2$ subject to $x_1^2 + x_2^2 + \cdots + x_N^2 = 1$.

(b) $x_1^2 + (x_1 + ax_2)^2 + \cdots + (x_1 + ax_2 + a^2 x_3 + \cdots + a^{N-1} x_N)^2$ subject to $x_1^2 + x_2^2 + \cdots + x_N^2 = 1$.

(c) $x_1^2 + (x_1 + ax_2)^2 + (-x_1 + ax_2 + (a+b)x_2)^2 + \cdots + (x_1 + ax_2 + (a+b)x_3 + \cdots + (a + (N-2)b)x_N)^2$ subject to $x_1^2 + x_2^2 + \cdots + x_N^2 = 1$.

30. Show that (5.9.3) can be written

$$f_c = \min_v \left[-\frac{(v^2 + c^2)}{v} \right].$$

Can one find an economic interpretation of this result?

31. Consider the problem of determining the minimum of

$$J(u) = \int_0^T (u'^4 + u^4)\, dt,$$

over all functions for which the integral exists and such that $u(0) = c$. Write $f(c, T) = \min J(u)$ and show that $f(c, T) = c^4 r(T)$. Show that $r'(T) = 1 - 3r^{4/3}$, $r(0) = 0$; show that $v(c, T)$ has the form $cg(T)$ and determine $g(T)$.

32. Determine the limiting form of $r(T)$ and $g(T)$ as $T \to \infty$.

33. If $J(u) = \int_0^T g(u', u)\, dt$, and $f(c, T) = \min_u J(u)$, where $u(0) = c$, show formally that $f_T = \min_v [g(c, v) + vf_c]$, $f(c, 0) = 0$.

34. If v has the form $h(T)c$, what are the possible forms for f and $g(c, v)$? (This is an important type of inverse problem, which we will discuss in Volume II in detail. See

R. Bellman and R. Kalaba, "An Inverse Problem in Dynamic Programming and Automatic Control," *J. Math. Anal. Appl.*, **7**, 1963, pp. 322–325.)

35. From the relation $r_N = (1 + a^2) - 1/(1 + r_{N-1})$, show how to deduce upper and lower bounds on the value of r_N. (*Hint:* Write

$$r_N = (1 + a^2) - \frac{a^2}{\left(1 + (1 + a^2) - \dfrac{a^2}{1 + r_{N-2}}\right)} .)$$

36. Consider the Euler equation associated with the minimization of $J(u) = \int_0^T (u'^2 + u^2)\, dt$, subject to $u(0) = 1$, namely $u'' - u = 0$, $u(0) = 1$, $u'(T) = 0$. Show that this equation has an unstable solution in the following senses:

    (a)   If $v'' - v = \varepsilon$, $v(0) = 1$, $v'(T) = 0$, then $|u - v|$ can become arbitrarily large as $T \to \infty$, for any $|\varepsilon| > 0$.

    (b)   If $v'' - v = 0$, $v(0) = 1 + \varepsilon_1$, $v'(T) = \varepsilon_2$, then $|u - v|$ can become arbitrarily large as $T \to \infty$, as long as $|\varepsilon_1| + |\varepsilon_2| > 0$.

37. Consider the Riccati equation associated with the foregoing minimization problem, $r' = 1 - r^2$, $r(0) = 0$. Show that this equation has a stable solution in the following senses:

    (a)   If $s' = 1 - s^2 + \varepsilon$, then $|s - r| \le k_1 \varepsilon$ for $T \ge 0$ for some constant k_1, provided that $|\varepsilon|$ is sufficiently small. Show this using the explicit solution and also using general techniques.

    (b)   If $s' = 1 - s^2 + \varepsilon(t)$, $s(0) = 0$, then $|s - r| \le k_1 \varepsilon$ for $T \ge 0$ provided that $|\varepsilon(t)| \le \varepsilon$ for $t \ge 0$, where ε is sufficiently small.

    (c)   If $s' = 1 - s^2$, $s(0) = \varepsilon$, then $|s - r| \le k_1 \varepsilon$ for $T \ge 0$ for some constant k_1, provided that $|\varepsilon|$ is sufficiently small.

(The point of the foregoing two exercises is to emphasize the fact that dynamic programming yields a stable computational approach whereas the approach of the calculus of variations automatically leads to an unstable equation.)

38. Consider the problem of minimizing the function $\sum_{n=0}^{N} (|u_n| + |v_n|)$ over $\{v_n\}$ with $u_{n+1} = au_n + v_n$, $u_0 = c$. Let $f_N(c)$ denote the minimum value. Show that

$$f_N(c) = \min_v \left[|c| + |v| + f_{N-1}(ac + v) \right],$$

$N \ge 1$, with $f_0(c) = |c|$. Hence show that $f_N(c) = r_N|c|$, where r_N is independent of c, and thus that

$$r_N = \min_v \left[1 + |v| + r_{N-1}(a + v) \right].$$

From this, obtain an explicit recurrence relation connecting r_{N-1} and r_N.

39. Consider the problem of minimizing $\max_{0 \le n \le N} |u_n| + \sum_{n=0}^{N} |v_n|$ over $\{v_n\}$ with $u_{n+1} = au_n + v_n$, $u_0 = c$. Consider the enlarged problem of minimizing $\max [b, \max_{0 \le n \le N} |u_n|] + \sum_{n=0}^{N} |v_n|$. Denote this minimum value by $f_N(b, c)$. Show that

$$f_N(b, c) = \min_v [|v| + f_{N-1}(\max (b, |c|), ac + v)],$$

$N \ge 1$, with $f_0(b, c) = \max (b, c)$, and thus that

$$f_N(b, c) = \min_v [|v| + f_{N-1}(|c|, ac + v)], \qquad 0 \le b \le |c|,$$

$$= \min_v [|v| + f_{N-1}(b, ac + v)], \qquad |c| \le b.$$

Can one deduce the analytic structure of $f_N(b, c)$ from this?

40. Consider the problem of minimizing $\max_{0 \le n \le N} |u_n| + \max_{0 \le n \le N} |v_n|$. (*Hint:* Consider the enlarged problem of minimizing the function

$$\max [b_1, \max_n |u_n|] + \max [b_2, \max |v_n|].)$$

41. Write $f(c, T) = \min J(u)$. Let $u(t, T)$ denote the minimizing function. Show that

$$f(c, T) \le \int_0^T [u'(t, T_1)^2 + u(t, T_1)^2] \, dt$$

for $T_1 \ge T$ and that

$$f(c, T_1) \le f(c, T) + (T - T_1)u^2(T, T).$$

(This is an application of a powerful technique called "transplantation." See

R. Bellman, "Functional Equations in the Theory of Dynamic Programming—VI: A Direct Convergence Proof," *Ann. of Math.*, **65**, 1957, pp. 215–223.

BIBLIOGRAPHY AND COMMENTS

5.1. For a careful analysis of the interconnections between dynamic programming and the calculus of variations see

S. Dreyfus, *Dynamic Programming and the Calculus of Variations*, Academic Press, New York, 1965.

5.2. Observe that we say "as if" not "is." It is sufficient for the mathematician that a control process may be viewed as if it were a multistage decision process.

We have omitted the difficulties involved in determining what time it is in a given state. The modern approach is to give equal attention to the errors connected with determining the position in phase space (state space) and the time.

5.3. The theory of dynamic programming was inaugurated at The RAND Corporation in 1949 initially in connection with some stochastic multistage decision processes of both one-person and two-person type. The first formal announcement was

R. Bellman, "The Theory of Dynamic Programming," *Proc. Nat. Acad. Sci. U.S.A.*, **38**, 1952, pp. 716–719,

and the first detailed exposition in

R. Bellman, *An Introduction to the Theory of Dynamic Programming*, The RAND Corporation, **R-245**, 1953.

R. Bellman, "The Theory of Dynamic Programming," *Bull. Amer. Math. Soc.*, **60**, 1954, pp. 503–516.

Subsequently, this was expanded to a book,

R. Bellman, *Dynamic Programming*, Princeton Univ. Press, Princeton, New Jersey, 1957.

Computational aspects were discussed in

R. Bellman and S. Dreyfus, *Applied Dynamic Programming*, Princeton Univ. Press, Princeton, New Jersey, 1962,

and the intimate connection with modern control theory is described in

R. Bellman, *Adaptive Control Processes: A Guided Tour*, Princeton Univ. Press, Princeton, New Jersey, 1961.

It is well to point out that "policy" is far more fundamental than optimization since it is meaningful to speak about policies in connection with processes containing vector-valued criteria or no quantitative criteria at all.

The concept of state and cause-and-effect is carefully analyzed in

L. A. Zadeh and C. A. Desoer, *Linear System Theory, The State Space Approach*, McGraw-Hill, New York, 1963.

Considerable physical intuition and mathematical ingenuity are required for the introduction of appropriate state variables. Some discussion of these matters is contained in

R. Bellman, R. Kalaba, and L. Zadeh, "Abstraction and Pattern Classification," *J. Math. Anal. Appl.*, **13**, 1966, pp. 1–7.

R. Bellman, "On the Application of Dynamic Programming to the Determination of Optimal Play in Chess and Checkers," *Proc. Nat. Acad. Sci. U.S.A.*, **53**, 1965, pp. 244–247.

R. Bellman, "Dynamic Programming, Generalized States, and Switching Systems," *J. Math. Anal. Appl.*, **12**, 1965, pp. 360–363.

R. Bellman, "Dynamic Programming, System Identification, and Suboptimization," *SIAM Control*, **4**, 1966.

R. Bellman, "Dynamic Programming, Pattern Recognition, and Location of Faults in Complex Systems," *J. Appl. Prob.*, **3**, 1966, pp. 268–271.

5.4. This technique was first presented in

R. Bellman, "Dynamic Programming and a New Formalism in the Calculus of Variations," *Proc. Nat. Acad. Sci. U.S.A.*, **40**, 1954, pp. 231–235.

R. Bellman, "Dynamic Programming and a New Formalism in the Calculus of Variations," *Rev. di Parma*, **6**, 1955, pp. 193–213.

5.7. The fact that the nonlinear partial differential equation could be solved by separation of variables was first indicated in

R. Bellman, "On a Class of Variational Problems," *Quart. Appl. Math.*, **14**, 1957, pp. 353–359.

See also

R. Bellman, "Functional Equations in the Theory of Dynamic Programming—VII: A Partial Differential Equation for the Fredholm Resolvent," *Proc. Amer. Math. Soc.*, **8**, 1957, pp. 435–440.

R. Bellman and H. Osborn, "Dynamic Programming and the Variation of Green's Functions," *J. Math. Mech.*, **7**, 1958, pp. 81–86.

R. Bellman, "Some New Techniques in the Dynamic Programming Solution of Variational Problems," *Quart. Appl. Math.*, **16**, 1958, pp. 295–305.

The multidimensional versions and stochastic versions of the results in the first reference were presented in Ph.D. theses by Adorno, Beckwith, and Freimer, and in papers by Kalman. References will be found at the end of Chapter 8.

5.11. Elimination of two-point boundary-value problems does not seem very important here in the one-dimensional case. It assumes a more dominant role in higher-dimensional control processes.

5.21. For further discussion of the use of these techniques, see

R. Bellman, I. Glicksberg, and O. Gross, *Some Aspects of the Mathematical Theory of Control Processes*, The RAND Corporation, **R-313**, 1958.

5.22. See

R. Bellman, "Functional Equations in the Theory of Dynamic Programming—VI: A Direct Convergence Proof," *Ann. of Math.*, **65**, 1957, pp. 215–223.

5.25. See

S. Dreyfus, *Dynamic Programming and the Calculus of Variations*, Academic Press, New York, 1965,

for a detailed discussion of the geometric background.

6

REVIEW OF MATRIX THEORY AND LINEAR

DIFFERENTIAL EQUATIONS

6.1. Introduction

Our aim is to study general control processes involving systems described by N state variables $x_1(t)$, $x_2(t)$, ..., $x_N(t)$ at time t. A basic problem worth examining for many reasons is that of minimizing the quadratic functional

$$J(x_1, x_2, \ldots, x_N; y_1, y_2, \ldots, y_N) = \int_0^T \left[\sum_{i=1}^{N} (x_i^2 + y_i^2) \right] dt \qquad (6.1.1)$$

over all admissible choices of $y_i(t)$, where the x_i and y_i are bound by the linear differential equations

$$\frac{dx_i}{dt} = \sum_{j=1}^{N} a_{ij} x_j + y_i, \qquad x_i(0) = c_i, \qquad i = 1, 2, \ldots, N. \qquad (6.1.2)$$

It is clear that unless we do something about notation, it is going to be very difficult to see what is occurring and to understand the structure of the optimal control and of the optimal policy. The only sensible way to handle multidimensional matters is by the use of vector-matrix notation. In this chapter, we shall briefly review what every aspiring control theorist should know about matrix theory and its applications to the theory of linear differential equations. What follows is not intended to be a first reading in this fundamental area of modern mathematics, although it is self-contained.

6.2. Vector-Matrix Notation

A column of numbers

$$x = \begin{pmatrix} x_1 \\ x_2 \\ \cdot \\ \cdot \\ \cdot \\ x_N \end{pmatrix} \tag{6.2.1}$$

will be called a vector. The quantities $x_1, x_2, \ldots, x_N$ are called the components. In the majority of cases of concern here, the x_i will be real. Two vectors x and y are equal if their respective components are equal, in which case we write $x = y$. The addition of two vectors is defined in the expected way,

$$x + y = \begin{pmatrix} x_1 + y_1 \\ x_2 + y_2 \\ \cdot \\ \cdot \\ \cdot \\ x_N + y_N \end{pmatrix}. \tag{6.2.2}$$

A one-dimensional vector, the usual complex number, will be called a scalar. The multiplication of a vector x by a scalar c_1 is defined by the relation

$$c_1 x = x c_1 = \begin{pmatrix} c_1 x_1 \\ c_1 x_2 \\ \cdot \\ \cdot \\ \cdot \\ c_1 x_N \end{pmatrix}. \tag{6.2.3}$$

A square array of numbers

$$A = \begin{pmatrix} a_{11} & a_{12} & \cdots & a_{1N} \\ a_{21} & a_{22} & \cdots & a_{2N} \\ \cdot & & & \\ \cdot & & & \\ \cdot & & & \\ a_{N1} & a_{N2} & \cdots & a_{NN} \end{pmatrix} = (a_{ij}) \tag{6.2.4}$$

is called a matrix. The quantities a_{ij} are called the elements of A. The addition of two matrices is also defined in the expected fashion, as is the multiplication of a matrix A by a scalar c_1,

$$A + B = (a_{ij} + b_{ij}),$$
$$Ac_1 = c_1 A = (c_1 a_{ij}). \tag{6.2.5}$$

The multiplication of a vector x by a matrix A is carefully defined so that the linear algebraic equation

$$\sum_{j=1}^{N} a_{ij} x_j = b_i, \qquad i = 1, 2, \ldots, N, \tag{6.2.6}$$

may be compactly written

$$Ax = b. \tag{6.2.7}$$

We see then that Ax is defined to be the vector whose ith component is $\sum_{j=1}^{N} a_{ij} x_j$. Note carefully the order of the terms. With this definition, it is easy to verify that

$$(A + B)x = Ax + Bx, \qquad A(x + y) = Ax + Ay. \tag{6.2.8}$$

EXERCISES

1. The vector 0 is defined as the vector with all of its components 0. It follows that $x + 0 = x$ for all x. Show that this defines 0 uniquely.
2. The matrix O is defined as the matrix with all of its components 0. It follows that $X + O = X$ for all X. Show that this defines O uniquely.
3. Show that $x + y + z$ and $X + Y + Z$ are unambiguously determined.

6.3. Inverse Matrix

If the system of linear algebraic equations of (6.2.5) can be written in the succinct form $Ax = b$, it is tempting to write the solution in the form $x = A^{-1}b$.

To justify this, let us begin by assuming that the determinant $|A| = |a_{ij}|, i, j = 1, 2, \ldots, N$, which we shall occasionally write as det (A), is nonzero. Then we know that (6.2.6) has a unique solution which can be obtained via Cramer's rule.

This solution has the form

$$x_i = \sum_{j=1}^{N} \alpha_{ij} b_j, \qquad i = 1, 2, \ldots, N, \qquad (6.3.1)$$

where α_{ij} is the cofactor of a_{ij} in the expansion of $|a_{ij}|$ divided by the value of the determinant. Let us then define

$$A^{-1} = (\alpha_{ij}). \qquad (6.3.2)$$

If $|A| = 0$, we call A singular. In this case, A^{-1} does not exist. It remains to show that A^{-1} has some of the usual properties of a reciprocal. To do this, we must introduce the concept of the product of two matrices.

EXERCISES

1. Show that if $|A| = 0$, there exists at least one nontrivial solution of $Ax = 0$. (*Hint:* Use induction.)

2. Show that A is nonsingular if $|a_{ii}| > \sum_{j \neq i} |a_{ij}|$ for $i = 1, 2, \ldots, N$.

(See the expository paper, O. T. Todd, "A Recurring Theorem on Determinants," *Amer. Math. Monthly*, **56**, 1949, pp. 672–676.)

6.4. The Product of Two Matrices

How to define AB? We are guided by the idea that a matrix is a symbol for a linear transformation. If $z = Ay$, $y = Bx$ denote respectively the linear transformations

$$z_i = \sum_{j=1}^{N} a_{ij} y_j, \qquad y_i = \sum_{j=1}^{N} b_{ij} x_j, \qquad i = 1, 2, \ldots, N, \qquad (6.4.1)$$

it is clear by direct substitution that the z_i are linear functions of the x_j. Since $z = A(Bx)$, it is natural to define AB as the matrix of this resultant linear transformation. Hence, we define the ijth element of AB to be $\sum_{k=1}^{N} a_{ik} b_{kj}$.

It is now easy to see that

$$(A + B)C = AC + BC, \qquad A(C + D) = AC + AD, \qquad (6.4.2)$$

and equally easy to see that in general

$$AB \neq BA. \qquad (6.4.3)$$

Hence, matrix multiplication is noncommutative. At first, this violation of the most important property of the ordinary multiplication appears to be a terrible nuisance. A little reflection will convince the reader that that it is actually an enormous boon. Many important transformations in the scientific domain are noncommutative, and matrix theory enables us to obtain a mathematical foothold. Furthermore, from the standpoint of the mathematician, life would be extremely boring without the challenge of noncommutativity.

A straightforward calculation shows that

$$(AB)C = A(BC), \qquad (6.4.4)$$

which means that ABC is an unambiguous expression. The simplest way to carry out this calculation is to employ the summation convention whereby a repeated index requires summation over this index. Thus,

$$a_{ik} b_{kj} = \sum_{k=1}^{N} a_{ik} b_{kj}. \qquad (6.4.5)$$

Hence,

$$(AB)C = ((a_{ik} b_{kl})c_{lj}),$$
$$A(BC) = (a_{ik}(b_{kl} c_{lj})), \qquad (6.4.6)$$

whence the usual associativity of multiplication yields the desired result.

It now follows readily from Cramer's rule that

$$AA^{-1} = I, \qquad (6.4.7)$$

where I, the identity matrix, is defined by

$$I = \begin{pmatrix} 1 & & & & 0 \\ & 1 & & & \\ & & \cdot & & \\ & & & \cdot & \\ 0 & & & & \cdot \\ & & & & & 1 \end{pmatrix}. \qquad (6.4.8)$$

EXERCISES

1. Prove that $A^{-1}A$ also equals I. In other words, despite the non-commutativity of multiplication, there is only one reciprocal of a matrix, which is simultaneously a left and right reciprocal.
2. If $AB = 0$, either A or B is singular. Must one or the other be equal to O?
3. Exhibit 2×2 matrices with the property that $AB \neq BA$.
4. Use the fact that $|AB| = |A| |B|$ to prove that $|A^{-1}| = |A|^{-1}$.
5. Use the fact that $|A|$ is the Jacobian of the transformation $y = Ax$ plus the geometric significance of the Jacobian to prove that $|AB| = |A| |B|$.
6. Show that the equation $XI = IX$ for all X determines I up to a scalar factor.
7. If Λ is the diagonal matrix

$$\Lambda = \begin{pmatrix} \lambda_1 & & & & 0 \\ & \lambda_2 & & & \\ & & \ddots & & \\ 0 & & & & \lambda_N \end{pmatrix},$$

and $\lambda_i \neq 0$, then

$$\Lambda^{-1} = \begin{pmatrix} \lambda_1^{-1} & & & & 0 \\ & \lambda_2^{-2} & & & \\ & & \ddots & & \\ 0 & & & & \lambda_N^{-1} \end{pmatrix}.$$

6.5. Inner Product and Norms

Let us now introduce an extremely important function of two vectors, the inner product,

$$(x, y) = \sum_{i=1}^{N} x_i y_i. \tag{6.5.1}$$

It is clear that $(x, y) = (y, x)$ and that $(x + y, z) = (x, z) + (y, z)$. If $(x, y) = 0$, we say that the vectors x and y are orthogonal. Furthermore, we see that

$$(Ax, y) = \sum_{i,j=1}^{N} a_{ij} x_i y_j, \tag{6.5.2}$$

and thus that

$$(Ax, y) = (x, A'y), \tag{6.5.3}$$

where A' is the transpose of A, the matrix (a_{ji}). For our further purposes the representation

$$(x, Ax) = \sum_{i,j=1}^{N} a_{ij} x_i x_j \tag{6.5.4}$$

is quite useful. It is clear that we may as well assume that $a_{ij} = a_{ji}$ if we are dealing with quadratic forms. A matrix with this property, $A' = A$, is called symmetric.

In the case that all of the components of x are real, the only case of interest to us in all that follows, the function $(x, x) = \sum_{i=1}^{N} x_i^2$, the square of the Euclidean distance, plays a useful role. It serves as the square of a norm $\|x\|$,

$$\|x\| = \left(\sum_{i=1}^{N} x_i^2 \right)^{1/2}. \tag{6.5.5}$$

It is not difficult to establish the triangle inequality

$$\|x + y\| \le \|x\| + \|y\|, \tag{6.5.6}$$

and the Cauchy-Schwarz inequality

$$(x, y)^2 \le (x, x)(y, y), \tag{6.5.7}$$

which asserts that the cosine of the angle between two vectors is at most 1 in absolute value.

EXERCISES

1. Show that A' is nonsingular if and only if A is nonsingular.
2. Show that $(x, Ax) = 0$ for all real x implies that $A' = -A$.
3. Hence, if A is symmetric, and $(x, Ax) = 0$, show that we must have $A = 0$.

4. If x and y are complex, define $[x, y] = (x, \bar{y})$, where $\bar{y}$ is the conjugate complex of y. Show that $[x, y]$ possesses the distributive properties of (x, y).

5. If the a_{ij} are real and $a_{11}r_1{}^2 + 2a_{12}r_1r_2 + a_{22}r_2{}^2 > 0$ for all non-trivial real r_1, r_2, then $a_{11}, a_{22} > 0$ and $a_{11}a_{22} - a_{12}^2 > 0$.

6. Using the quadratic form $(r_1x + r_2y, r_1x + r_2y)$ show that $(x, x)(y, y) \geq (x, y)^2$ for all real x and y with strict inequality if y is not proportional to x.

7. Use the relation $(x + y, x + y) = (x, x + y) + (y, x + y)$ together with the Cauchy-Schwarz inequality to establish the triangle inequality.

8. Show that $(x, x)^{1/2} = \max_y (x, y)$, where y ranges over $(y, y) = 1$, and thus establish the triangle inequality.

9. If a_{ij} are real and $\sum_{i,j=1}^3 a_{ij}r_ir_j > 0$ for all nontrivial real r_1, r_2, then

$$a_{11} > 0, \qquad \begin{vmatrix} a_{11} & a_{12} \\ a_{12} & a_{22} \end{vmatrix} > 0, \qquad \begin{vmatrix} a_{11} & a_{12} & a_{13} \\ a_{12} & a_{22} & a_{23} \\ a_{13} & a_{23} & a_{33} \end{vmatrix} > 0.$$

This is a necessary and sufficient condition for the positive definiteness of the quadratic form $\sum a_{ij}r_ir_j$.

10. Show that $(AB)' = B'A'$ and thus that $A'A$ is a symmetric matrix.

6.6. Orthogonal Matrices

A matrix which preserves Euclidean distance is called orthogonal. If $\|Tx\| = \|x\|$, we must have

$$(Tx, Tx) = (x, x), \qquad (6.6.1)$$

whence $(x, T'Tx) = (x, x)$, for all x. Hence, since $T'T$ is symmetric, we must have $T'T = I$. From this we see that $T^{-1} = T'$ for an orthogonal matrix.

EXERCISE

1. Show that orthogonal matrices form a group; that is, that the product of two orthogonal matrices is again orthogonal, that every orthogonal matrix has an inverse that is orthogonal, and that the identity matrix is an orthogonal matrix. The associativity property has already been established.

6.7. Canonical Representation

The fundamental theorem concerning symmetric matrices, upon which all else can be made to depend, is the following:

Theorem. A real symmetric matrix A can be transformed into a diagonal matrix by means of an orthogonal transformation

$$A = T \begin{pmatrix} \lambda_1 & & & & 0 \\ & \lambda_2 & & & \\ & & \cdot & & \\ & & & \cdot & \\ 0 & & & & \cdot \\ & & & & \lambda_N \end{pmatrix} T'. \qquad (6.7.1)$$

The proof proceeds along the following lines. Let us look for the vectors that are invariant in direction under the transformation A, namely those satisfying the equation

$$Ax = \lambda x, \qquad (6.7.2)$$

where λ is a scalar. Since this is a homogeneous system, we see that λ must satisfy the determinantal equation

$$|A - \lambda I| = 0. \qquad (6.7.3)$$

Let $\lambda_1, \lambda_2, \ldots, \lambda_N$ be the N roots of this equation distinct or not, the characteristic roots of A. Then it can be shown that A has a full set of characteristic vectors that can be taken to be mutually orthogonal and to have norm 1. Let $x^{(1)}, x^{(2)}, \ldots, x^{(N)}$ denote these vectors. Then, set

$$T = \begin{pmatrix} x^{(1)} & x^{(2)} & \cdots & x^{(N)} \end{pmatrix}, \qquad (6.7.4)$$

which is to say T is the matrix whose N columns are the $x^{(i)}$. From what we have said, we see that T is orthogonal, $T'T = I$. The N equations

$$Ax^{(i)} = \lambda_i x^{(i)} \qquad (6.7.5)$$

yield the matrix equation

$$AT = T\Lambda, \qquad (6.7.6)$$

where Λ is the diagonal matrix of (6.7.1). Hence,

$$A = T\Lambda T^{-1} = T\Lambda T',\qquad (6.7.7)$$

the desired representation.

EXERCISES

1. Show that if $x = Ty$, then $(x, Ax) = (y, \Lambda y) = \sum_{i=1}^{N} \lambda_i y_i^2$.
2. Show that $\lambda_N(x, x) \geq (x, Ax) \geq \lambda_1(x, x)$ if $\lambda_1 \leq \lambda_2 \leq \cdots \leq \lambda_N$.
3. Show that A^{-1} has the characteristic vectors λ_i^{-1} if A is non-singular and that $A^{-1} = T\Lambda^{-1}T'$.

6.8. Det A

We see from (6.7.7) that

$$|A| = |T\Lambda T'| = |T|\,|\Lambda|\,|T'| = |\Lambda|\,|T|\,|T'| = |\Lambda|. \qquad (6.8.1)$$

Since $T'T = I$, we see that $|T'|\,|T| = 1$.

EXERCISES

1. Using (6.8.1), show $|A| = \prod_{i=1}^{N} \lambda_i$.
2. Derive this directly from (6.7.3).
3. Define tr $(A) = \sum_{i=1}^{N} a_{ii}$. This is called the trace of A. Show that tr $(A) = \sum_{i=1}^{N} \lambda_i$.
4. Show that tr $(AB) = $ tr (BA).

6.9. Functions of a Symmetric Matrix

We have already considered the reciprocal matrix, A^{-1}. For our further purposes, we need the exponential matrix e^A. Let us begin with the case where A is real and symmetric. As we shall see below, there are several ways of defining this matrix. From the representation of (6.7.7), we see that

$$A^2 = (T\Lambda T')(T\Lambda T') = T\Lambda(T'T)\Lambda T' = T\Lambda^2 T', \qquad (6.9.1)$$

and inductively that $A^k = T\Lambda^k T'$, $k = 1, 2, \ldots$. Hence, by analogy with the usual exponential series, it is reasonable to set

$$
e^A = \sum_{k=0}^{\infty} \frac{A^k}{k!} = T \begin{pmatrix} e^{\lambda_1} & & & & 0 \\ & e^{\lambda_2} & & & \\ & & \cdot & & \\ & & & \cdot & \\ 0 & & & & \cdot \\ & & & & & e^{\lambda_N} \end{pmatrix} T', \qquad (6.9.2)
$$

where $\lambda_1, \lambda_2, \ldots, \lambda_N$ are the characteristic roots in some fixed order. At the moment, we use the first and third terms of (6.9.2) for our definition of e^A, since we have not discussed the convergence of the infinite series.

It follows that

$$
e^{-A} = T \begin{pmatrix} e^{-\lambda_1} & & & & 0 \\ & e^{-\lambda_2} & & & \\ & & \cdot & & \\ 0 & & & \cdot & \\ & & & & \cdot \\ & & & & & e^{-\lambda_N} \end{pmatrix} T', \qquad (6.9.3)
$$

since the characteristic roots of $-A$ are $-\lambda_1, -\lambda_2, \ldots, -\lambda_N$. Thus,

$$
e^A e^{-A} = T \begin{pmatrix} e^{\lambda_1} & & \\ & \cdot & \\ & & e^{\lambda_N} \end{pmatrix} T'T \begin{pmatrix} e^{-\lambda_1} & & \\ & e^{-\lambda_2} & \\ & & \cdot \\ & & & e^{-\lambda_N} \end{pmatrix} T = I. \qquad (6.9.4)
$$

Hence, e^{-A} is the reciprocal of e^A, and what is most important, this shows that e^A is never singular.

Since the characteristic roots of e^{At} are $\lambda_1 t, \lambda_2 t, \ldots, \lambda_N t$, for t a scalar, we see that

$$
e^{At} = T \begin{pmatrix} e^{\lambda_1 t} & & & & 0 \\ & e^{\lambda_2 t} & & & \\ & & \cdot & & \\ 0 & & & \cdot & \\ & & & & \cdot \\ & & & & & e^{\lambda_N t} \end{pmatrix} T'. \qquad (6.9.5)
$$

Thus,

$$e^{A(t+s)} = e^{At}e^{As} \tag{6.9.6}$$

for s, t any two scalars. So far we have established this only for the case where A is symmetric. Actually, as we see below, it holds in general.

EXERCISE

1. Show that $e^{(A+B)t} = e^{At}e^{Bt}$ for all t if and only if $AB = BA$.

6.10. Positive Definite Matrices

If A is a real symmetric matrix, the quadratic form

$$(x, Ax) = \sum_{i,j=1}^{N} a_{ij} x_i x_j \tag{6.10.1}$$

in general will assume both positive and negative values as the x_i range over all real values. If it is the case that

$$(x, Ax) > 0 \tag{6.10.2}$$

apart from the obvious exceptional set $x_1 = x_2 = \cdots = x_N = 0$ (the null vector), we say that A is positive definite, and write $A > 0$. If A is merely nonnegative, that is, if $(x, Ax) \geq 0$ for all nontrivial x, we write $A \geq 0$.

From the canonical form of Section 6.7, we see that a necessary and sufficient condition that A be positive definite is that all of its characteristic roots be positive.

If A and B are two symmetric matrices and $A - B > 0$, we write $A > B$. This inequality relation possesses certain nonintuitive features. Namely, $A > B$ does not necessarily imply $A^2 > B^2$. It is, however, true that $A > B > 0$ implies $B^{-1} > A^{-1} > 0$, as we demonstrate in the following section.

EXERCISES

1. $A \geq B$, $B \geq C$ implies $A \geq C$.
2. $A \geq B$ implies that $CAC' \geq CBC'$ for any C.
3. Exhibit 2×2 matrices such that $A > B$ does not imply $A^2 > B^2$.

4. Show that the maximum of $2(x, y) - (y, Ay)$, with respect to y, where $A > 0$, is attained at $y = Ax$.

5. Define the square root of A as follows:

$$A^{1/2} = T \begin{pmatrix} \lambda_1^{1/2} & & & & 0 \\ & \lambda_2^{1/2} & & & \\ & & \cdot & & \\ & & & \cdot & \\ 0 & & & & \lambda_N^{1/2} \end{pmatrix} T'$$

for a positive definite A. Show that A has a unique positive definite square root. How many symmetric square roots does it have?

6. Define $\cosh A = (e^A + e^{-A})/2$. Show that $\cosh A$ is never singular.

6.11. Representation of A^{-1}

To establish the desired result, we employ the representation

$$(c, A^{-1}c) = \max_x [2(x, c) - (x, Ax)] \qquad (6.11.1)$$

for $A > 0$. We see upon differentiation that the maximizing vector satisfies the equation $Ax = c$, whence (6.11.1) follows.

From (6.11.1), we obtain the stated relation since

$$(c, B^{-1}c) = \max_x [2(x, c) - (x, Bx)]$$

$$\geq \max_x [2(x, c) - (x, Ax)] = (c, A^{-1}c) \qquad (6.11.2)$$

if $A > B$.

EXERCISE

1. This establishes $B^{-1} \geq A^{-1}$ if $A > B > 0$. Show that the inequality is strict by modifying the foregoing argument. (*Hint:* $A > B$ implies $A > B + \varepsilon I$ for some positive ε.)

6.12. Differentiation and Integration of Vectors and Matrices

Let us now consider the case where the components of x depend on a real variable. We will write $x(t)$ and define differentiation and integration in the expected fashion.

$$\frac{dx}{dt} = \begin{pmatrix} \dfrac{dx_1}{dt} \\ \vdots \\ \dfrac{dx_N}{dt} \end{pmatrix}, \qquad \int x \, dt = \begin{pmatrix} \int x_1 \, dt \\ \vdots \\ \int x_N \, dt \end{pmatrix}. \qquad (6.12.1)$$

With this notation, we can write the linear system of differential equations

$$\frac{dx_i}{dt} = \sum_{j=1}^{N} a_{ij} x_j, \qquad x_i(0) = c_i, \qquad i = 1, 2, \ldots, N, \qquad (6.12.2)$$

in the form

$$\frac{dx}{dt} = Ax, \qquad x(0) = c. \qquad (6.12.3)$$

Similarly, we define differentiation and integration of matrices. What is remarkable is that it is easiest to discuss (6.12.3) in terms of the solution of the matrix differential equation

$$\frac{dX}{dt} = AX, \qquad X(0) = I. \qquad (6.12.4)$$

EXERCISES

1. Show that the definitions of dx/dt and $\int x \, dt$ are forced upon us as a consequence of the limiting forms of the sum and difference of two vectors.

2. Show that $d/dt(XY) = dX/dt \ Y + X \ dY/dt$ and that $d/dt(x, y) = (dx/dt, y) + (x, dy/dt)$.

3. Show that

$$\frac{d}{dt}(X^2) = X\frac{dX}{dt} + \left(\frac{dX}{dt}\right)X,$$

$$\frac{d}{dt}(X^{-1}) = -X^{-1}\frac{dX}{dt}X^{-1},$$

and that

$$\frac{d}{dt}(e^X) = e^X\frac{dX}{dt}$$

if X and dX/dt commute. Is the relation true if X and dX/dt do not commute?

4. Show how to obtain a first-order system equivalent to the Nth order linear differential equation $u^{(N)} + p_1 u^{(N-1)} + \cdots + p_N u = 0$. (*Hint:* Set $u = x_1, u' = x_2, \ldots, u^{(N)-1} = x_N$.)

6.13. The Matrix Exponential

If A is real and symmetric, we know that the matrix e^{At} is a solution of (6.12.4). Let us now show that we can define a matrix exponential e^{At} for all square matrices by means of the unique solution of the equation

$$\frac{dX}{dt} = AX, \qquad X(0) = I. \tag{6.13.1}$$

This matrix function will also satisfy the functional equation

$$e^{A(t+s)} = e^{At}e^{As} \tag{6.13.2}$$

for all scalar s and t, and thus $(e^A)^{-1} = e^{-A}$. Hence, this matrix function is never singular.

6.14. Existence and Uniqueness Proof

In order to make the text reasonably self-contained, let us briefly sketch an existence and uniqueness proof for the solution of not only (6.12.4), but the more general equation

$$\frac{dX}{dt} = A(t)X, \qquad X(0) = I, \tag{6.14.1}$$

where $A(t)$ is a continuous function of t for $t \geq 0$. We wish to establish

Theorem. *If $A(t)$ is continuous for $t \geq 0$, (6.14.1) has a unique solution that is never singular for $t \geq 0$.*

To establish the existence, we employ successive approximations. Let $\{X_n\}$ be defined recurrently,

$$\frac{dX_{n+1}}{dt} = A(t)X_n, \qquad X_{n+1}(0) = I, \tag{6.14.2}$$

$n = 0, 1, \ldots$, with $X_0 = I$. Introduce the norm of a matrix

$$\|B\| = \sum_{i,j=1}^{N} |b_{ij}|. \tag{6.14.3}$$

Then it is easy to see that

$$\begin{aligned}
\|B + C\| &\leq \|B\| + \|C\|, \\
\|BC\| &\leq \|B\| \, \|C\|, \\
\left\| \int_0^t X \, dt_1 \right\| &\leq \int_0^t \|X\| \, dt_1.
\end{aligned} \tag{6.14.4}$$

From (6.14.2) we have

$$X_{n+1} = I + \int_0^t A(t_1)X_n \, dt_1,$$

$$X_{n+1} - X_n = \int_0^t A(t_1)(X_n - X_{n-1}) \, dt_1. \tag{6.14.5}$$

Hence, for $n > 1$,

$$\|X_{n+1} - X_n\| \leq \int_0^t \|A(t_1)\| \, \|X_n - X_{n-1}\| \, dt_1, \tag{6.14.6}$$

and thus inductively for $n \geq 0$,

$$\|X_{n+1} - X_n\| \leq \frac{\left(\int_0^t \|A(t_1)\| \, dt_1 \right)^{n+1}}{(n+1)!}. \tag{6.14.7}$$

Hence, the matrix series converges uniformly being majorized term-by-term by the series

$$\sum_{n=1}^{\infty} \frac{\left(\int_0^t \|A(t_1)\| \, dt_1 \right)^n}{n!}.$$

To establish uniqueness, we suppose that Y is another solution. Then

$$X = I + \int_0^t A(t_1)X \, dt_1,$$

$$Y = I + \int_0^t A(t_1)Y \, dt_1, \qquad (6.14.8)$$

and thus

$$X - Y = \int_0^t A(t_1)(X - Y) \, dt_1,$$

$$\|X - Y\| \leq \int_0^t \|A(t_1)\| \, \|X - Y\| \, dt_1$$

$$< \varepsilon + \int_0^t \|A(t_1)\| \, \|X - Y\| \, dt_1 \qquad (6.14.9)$$

for any $\varepsilon > 0$. Using Exercise 5 of Section 2.5, this implies that

$$\|X - Y\| \leq \varepsilon \exp\left(\int_0^t \|A(t_1)\| \, dt_1\right). \qquad (6.14.10)$$

Since this holds for any $\varepsilon > 0$, we must have $\|X - Y\| = 0$. Hence, $X = Y$.

We maintain that $x = Xc$ is a solution of the vector equation

$$\frac{dx}{dt} = A(t)x, \qquad x(0) = c, \qquad (6.14.11)$$

and the same proof as above shows that it is the unique solution.

To show that $X(t)$ is nonsingular, we can proceed in several ways. To begin with, we can establish the explicit representation

$$|X(t)| = \exp\left(\int_0^t \text{tr} \, (A(t_1)) \, dt_1\right), \qquad (6.14.12)$$

where $\text{tr} \, (A) = \sum_{i=1}^N a_{ii}$, the sum of the elements along the main diagonal. We leave this formula of Jacobi as an exercise for the reader.

Let us, however, employ a different technique, which we shall find quite handy subsequently. Suppose that $X(t_1)$ is singular. Then a non-trivial vector b exists such that

$$X(t_1)b = 0. \qquad (6.14.13)$$

Hence $y = X(t)b$ is a solution of

$$\frac{dy}{dt} = A(t)y, \qquad y(t_1) = 0. \qquad (6.14.14)$$

But the preceding existence and uniqueness shows that $y \equiv 0$ for $t \geq 0$. This contradicts the fact that $y(0) = Ib = b$, which is not identically zero.

EXERCISES

1. Show that

$$X = \exp\left(\int_0^t A(t_1)\, dt_1\right)$$

is the solution of (6.14.1) if and only if

$$A(t)\int_0^t A(t_1)\, dt_1 = \left(\int_0^t A(t_1)\, dt_1\right)A(t)$$

for $t \geq 0$. Exhibit a matrix $A(t)$ for which this condition is not satisfied.

2. Use the foregoing techniques to establish the convergence of the series $\sum_{n=1}^{\infty} A^n t^n / n!$ for all complex t. Show that this series satisfies the differential equation $dX/dt = AX$, $X(0) = I$. Hence, show that it is equal to e^{At}.

3. Use the series to show that $e^{A(t+s)} = e^{At}e^{As}$ for any constant matrix A.

4. Show that the solution of $dx/dt = Ax + f$, $x(0) = c$, may be written

$$x = e^{At}c + \int_0^t e^{A(t-t_1)}f(t_1)\, dt_1.$$

(*Hint:* Use the integrating factor e^{-At}.)

5. Show that the solution of $dx/dt = A(t)x + f$, $x(0) = c$, may be written

$$x = X(t)c + \int_0^t X(t)X(t_1)^{-1}f(t_1)\,dt_1,$$

where $X(t)$ is the solution of $dX/dt = A(t)X$, $X(0) = I$.

6.15. Euler Technique and Asymptotic Behavior

An alternative approach to the determination of the solution of

$$\frac{dx}{dt} = Ax, \qquad x(0) = c, \tag{6.15.1}$$

is to find particular solutions of the form $x = e^{\lambda t}b$, where b is a constant vector, and then to use superposition. Substituting in (6.15.1), we see that b is a characteristic vector of A and λ is a characteristic root.

The theory is much more difficult for general A than for symmetric A since A need not possess a set of linear independent characteristic vectors. Nonetheless, there are procedures that overcome this defect such as the use of the Jordan canonical form or the use of the semidiagonal form. We shall not enter into these matters here since we will not have occasion to use them.

What is important to point out is that the asymptotic behavior of $x(t)$ as $t \to \infty$ is determined by the characteristic root of A, λ_1, with largest real part. If λ_1 is simple, we have

$$x(t) \sim e^{\lambda_1 t}b_1; \tag{6.15.2}$$

if λ_1 is a multiple root, powers of t may appear.

From this, we can see that the long-term behavior of quite complicated systems can be described very simply. The fact that (6.15.2) holds plays an essential role in dealing with e^{At} for large t. Although never singular, it behaves more and more like an ill-conditioned matrix. We will discuss this in Chapter 7.

EXERCISES

1. Consider the system $x_1' = x_1$, $x_2' = x_2 + x_1$, $x_1(0) = x_2(0) = 1$ to show that the term te^t enters.
2. Construct an Nth order system in which the term t^{N-1} enters as a coefficient of the exponential.

6.16. $x'' - A(t)x = 0$

Let us now consider the vector differential equation

$$x'' - A(t)x = 0 \tag{6.16.1}$$

subject to the two-point boundary conditions

$$x(0) = c, \qquad x'(T) = 0, \tag{6.16.2}$$

an equation which arises, as we shall see, as the Euler equation associated with a multidimensional variational problem.

Consider first the matrix differential equation

$$X'' - A(t)X = 0, \tag{6.16.3}$$

and let $X_1(t)$ and $X_2(t)$ be the principal solutions, which is to say X_1 and X_2 satisfy (6.16.3), and, respectively, the initial conditions

$$X_1(0) = I, \quad X_1'(0) = 0; \qquad X_2(0) = 0, \quad X_2'(0) = I. \tag{6.16.4}$$

Set

$$x = X_1(t)c^{(1)} + X_2(t)c^{(2)}, \tag{6.16.5}$$

where $c^{(1)}$ and $c^{(2)}$ are constant vectors to be determined. Using the condition at $t = 0$, we see that $c^{(1)} = c$. The condition at $t = T$ yields the equation

$$0 = X_1'(T)c + X_2'(T)c^{(2)}. \tag{6.16.6}$$

Hence, if $X_2'(T)$ is nonsingular, we obtain a unique solution

$$c^{(2)} = -X_2'(T)^{-1}X_1'(T)c \tag{6.16.7}$$

and thus a unique solution to (6.16.3). In the cases of interest to us, $X_2'(T)$ will be nonsingular.

More general boundary conditions are dealt with in a similar manner. But the discussion of existence and uniqueness of solution can become quite complex. Fortunately, those that arise in control theory can be handled in a reasonably straightforward way.

6.17. $x' = Ax + By, \quad y' = Cx + Dy$

The general linear system

$$x' = Ax + By, \qquad y' = Cx + Dy, \tag{6.17.1}$$

is dealt with in an analogous fashion. We consider first the matrix system

$$X' = AX + BY, \qquad Y' = CX + DY, \tag{6.17.2}$$

and the principal solutions

$$\begin{aligned} X_1(0) &= I, & Y_1(0) &= 0, \\ X_2(0) &= 0, & Y_2(0) &= I. \end{aligned} \tag{6.17.3}$$

Then the general solution of (6.17.1) has the form

$$\begin{aligned} x &= X_1 c^{(1)} + X_2 c^{(2)}, \\ y &= Y_1 c^{(1)} + Y_2 c^{(2)}, \end{aligned} \tag{6.17.4}$$

where $c^{(1)}$ and $c^{(2)}$ are arbitrary constant vectors.

The solution of (6.17.1) subject to conditions such as

$$x(0) = c, \qquad y(T) = 0, \tag{6.17.5}$$

reduces to solving the system

$$\begin{aligned} c &= c^{(1)}, \\ 0 &= Y_1(T)c^{(1)} + Y_2(T)c^{(2)}. \end{aligned} \tag{6.17.6}$$

We see again that there is a unique solution if $Y_2(T)$ is nonsingular.

6.18. Matrix Riccati Equation

Let X and Y be two solutions of (6.17.2) and consider the matrix $Z = XY^{-1}$. We have

$$\frac{dZ}{dt} = \frac{dX}{dt} Y^{-1} - XY^{-1}\frac{dY}{dt} Y^{-1}$$
$$= (AX + BY)Y^{-1} - XY^{-1}(CX + DY)Y^{-1}$$
$$= AXY^{-1} + B - XY^{-1}CXY^{-1} - XY^{-1}D$$
$$= AZ + B - ZCZ - ZD, \qquad (6.18.1)$$

a matrix Riccati equation. Equations of this type play an essential role in the theories of dynamic programming and invariant imbedding. We will discuss (6.18.1) again in Chapter 8.

EXERCISE

1. Show that Z^{-1} is a solution of a Riccati equation when Z is. Show that $(C_1Z + C_2)(C_3Z + C_4)^{-1}$ is a solution of a Riccati equation when Z is. How can one obtain this result without calculation?

6.19. $dX/dt = AX + XB$, $X(0) = C$

Finally, let us note that the solution of the linear approximation to (6.18.1),

$$\frac{dX}{dt} = AX + XB, \qquad X(0) = C \qquad (6.19.1)$$

is

$$X = e^{At}Ce^{Bt}. \qquad (6.19.2)$$

EXERCISES

1. What is the solution of $dX/dt = AX + XB + F(t)$, $X(0) = C$? (*Hint:* Use two integrating factors.)

2. If the characteristic roots of A and B have negative real parts, the solution of $AX + XB = C$ is given by $X = \int_0^\infty e^{At} C e^{Bt}\, dt$. (*Hint:* That it is *a* solution follows from (6.19.1); that it is *the* solution follows from the linearity of the equation.)

3. Show that the solution of $dX/dt = AX + XA'$, $X(0) = C$, is symmetric if C is symmetric.

4. Show that if $dX/dt \geq AX + XA'$ for $t \geq 0$, with $X(0) = 0$, then $X(t) \geq 0$ for $t \geq 0$.

Miscellaneous Exercises

1. Let $f(c) = f(c_1, c_2, \ldots, c_N)$ be a scalar function of N variables, and grad f denote the vector whose ith component is $\partial f/\partial c_i$,

$$\text{grad } f = \begin{pmatrix} \partial f/\partial c_1 \\ \vdots \\ \partial f/\partial c_N \end{pmatrix}.$$

2. Show that the multidimensional Taylor expansion may be written $f(c + b) = f(c) + (b, \text{grad } f) + \cdots$. Obtain an expression for the next term.

3. Let $\{A_n\}$ be a sequence of positive definite matrices such that $A_1 \leq A_2 \leq \cdots \leq A_n \leq \cdots \leq B$. Show that A_n converges. (*Hint:* $(c, A_n c)$ is a monotone increasing scalar function for all c. Choose c adroitly.)

4. If A is symmetric and B is positive definite, then $AB + BA = 0$ implies that $A = 0$.

5. If B is positive definite, $AB + BA = C$ and C is positive definite, then A is positive definite.

6. Define $X = [C, B]$ to be the unique solution of $XB + BX = 2C$, provided that no characteristic root is equal to zero and the sum of no two characteristic roots is zero. Show that $[C, [C, B]] = B$.

7. Show that A, B positive definite and $A^2 > B^2$ implies $A > B$ using the identity $A^2 - B^2 = \frac{1}{2}(A + B)(A - B) + \frac{1}{2}(A - B)(A + B)$.

(For proofs of these results, together with applications, see

E. P. Wigner and M. M. Yanase, "On the Positive Semidefinite Nature of a Certain Matrix Expression," to appear.)

8. If B is positive definite with no characteristic value in $[0, k]$, then $B - kI$ is positive definite.
   (See

   G. Temple, "An Elementary Proof of Kato's Lemma," *Mathematika*, **2**, 1955, pp. 39–41.)

9. If A is a symmetric matrix that has no characteristic value in the closed interval $[a, b]$, then $(A - aI)(A - bI)$ is positive definite. (Kato's lemma.)

10. Show that $AX - XA = I$ has no solution. (*Hint:* Consider the effect of taking the trace of both sides.)

11. Consider the equation $x' = Ax$, $x(0) = c$, where A and c are real. Show that if Y is chosen so that $A'Y + YA = -I$, then $d/dt(x, Yx) = -(x, x)$.

12. Suppose that A is a stability matrix and we wish to calculate $J = \int_0^\infty (x, Bx)\, dt$, where $x' = Ax$, $x(0) = c$. Then $J = -(c, Yc)$, where Y is the solution of $B = A'Y + YA$.

13. Show that a necessary and sufficient condition that a real matrix A be a stability matrix is that the solution of $A'Y + YA = -I$ be positive definite. (*Hint:* $\int_0^T (x, x)\, dt + (x(T), Yx(T)) = (x(0), yx(0))$.)

14. Show that if A, B, and C are nonnegative definite and either A or C is positive definite, then $|\lambda^2 A + 2\lambda B + C| = 0$ has no roots with positive real parts. If A and C are nonnegative definite and B is positive definite, then the only root with zero real part is $\lambda = 0$. (*Hint:* Start with the equation $Ax'' + 2Bx' + Cx = 0$ and consider the integral $\int_0^T (x', Ax'' + 2Bx' + Cx)\, dt = 0$.)

15. Suppose that C is positive definite. Show by direct calculation that $\min_x \max_y f(x, y) = \max_y \min_x f(x, y)$, where

$$f(x, y) = (x, Cx) + 2(x, By) - (y, Cy) - 2(b, x) - 2(a, y).$$

   Show that the variational equations are $Bx - Cy = a$, $Cx + By = b$.

16. Show that for this function $f(x, y)$ the minimum with respect to the components of x and the maximum with respect to the components of y can be taken in any order; for example,

$$\min_{x_i} \max_{y_j} = \min_{x_1} \max_{y_1} \min_{[x_2,\ldots,x_N]} \max_{[y_2,\ldots,y_N]} .$$

   Is this true for any function of the vectors x and y? Consider the scalar case $f(x_1, y_1)$.

17. Let A be an M-dimensional matrix and B an N-dimensional matrix. The MN-dimensional matrix defined by $C = (a_{ij}B)$ is called the Kronecker product of A and B and written $A \times B$. Show that the characteristic roots of $A \times B$ are $\lambda_i \mu_j$, where λ_i are the characteristic roots of A and μ_j the characteristic roots of B. (*Hint:* Begin with the 2×2 case to simplify the algebra and consider the equation obtained for the $x_i y_j$ upon multiplying in all possible ways the equations equivalent to $\lambda_i x = Ax$, $\mu_j y + By$.)

18. Show that
$$A \times B \times C = (A \times B) \times C = A \times (B \times C),$$
$$(A + B) \times (C + D) = A \times C + A \times D + B \times C + B \times D,$$
$$(A \times B)(C \times D) = (AC) \times (BD).$$

19. Let x satisfy $dx/dt = Ax$, y satisfy $dy/dt = By$. Form the differential equation of order N^2 satisfied by the N^2 terms $x_i y_j$, $i, j = 1, 2, \ldots, N$. Show that the matrix of coefficients may be written $C = A \times I + I \times B$, and that the characteristic roots are $\lambda_i + \mu_j$.

20. Consider the equation $AX + XB = C$. Considering the elements of X, x_{ij}, as components of an N^2-dimensional vector, show that this matrix equation can be written as a linear system whose matrix is $A \times I + I \times B'$.

21. Hence show that a necessary and sufficient condition that $AX + XB = C$ be soluble for all C is that $\lambda_i + \mu_j \neq 0$ for any i and j.

22. Let Y be determined by the relation $A'Y + YA = -I$, where A is real and A' is, as above, the transpose. Then a necessary and sufficient condition that A be a stability matrix is that Y be positive definite.

23. Obtain a perturbation series
$$e^{A + \varepsilon B} = e^A + \varepsilon f_1(A, B) + \varepsilon^2 f_2(A, B) + \cdots,$$
by considering $e^{A + \varepsilon B}$ to be $X(1)$, where $X' = (A + \varepsilon B)X$, $X(0) = I$. (*Hint:* Convert the differential equation into an integral equation
$$X = e^{At} + \varepsilon \int_0^t e^{A(t - t_1)} BX(t_1) \, dt_1.)$$

24. Show that $dx/dt = A(t)x$, $x(0) = c$ with the components of c positive implies that the components of $x(t)$ are positive for $t \geq 0$ if $a_{ij}(t) \geq 0$, $i \neq j$ for $t \geq 0$.

25. Show that if $A(t)$ is a constant matrix, then $a_{ij} \geq 0$, $i \neq j$ is a necessary and sufficient condition.

26. Show that $X(t + s) = X(t)X(s)$ for $-\infty < s, t < \infty$ and continuity of $X(t)$ implies that $X(t) = e^{At}$ for some constant A.

27. If A is symmetric and positive definite with $A \leq I$ then the sequence $\{X_n\}$ defined by $X_{n+1} = X_n + (A - X_n^2)/2$, $X_0 = 0$, converges to the positive definite square root of A.

28. Show that the solution of the linear difference equation $x(n + 1) = Ax(n)$, $x(0) = c$, is given by $x(n) = A^n c$. Obtain a representation for the solution of $x(n + 1) = Ax(n) + g(n)$, $x(0) = c$.

29. What conditions must be imposed in order that $\lambda^n c$ be a solution of $x(n + 1) = Ax(n)$?

30. Show that the general solution of $x(n + 1) + Ax(n)$ has the form $x(n) = \sum_{i=1}^{N} p_i(n)\lambda_i^n$, where the $p_i(n)$ are vectors whose components are polynomials of degree at most $N - 1$. If the λ_i are all different, the $p_i(n)$ are constant.

31. What is a necessary and sufficient condition that all solutions of $x(n + 1) = Ax(n)$ approach zero as $n \to \infty$?

32. Show that $\lim_{n \to \infty} (1 + A/n)^n = e^A$.

33. Let $x((n + 1)\Delta) = x(n\Delta) + A \, \Delta x(n\Delta)$, $x(0) = c$, with $\Delta > 0$ a scalar. Show that, as $\Delta \to 0$, $x(n\Delta) \to x(t)$, where $x'(t) = Ax(t)$, $x(0) = c$, provided that $n\Delta \to t$.

34. Given any matrix A we can find a matrix $B(\varepsilon)$ with distinct characteristic roots such that $\|A - B\| \leq \varepsilon$ for any positive ε.

35. Show that a nonsingular matrix A may be written as an exponential, $A = e^B$.

36. Show that the solution of $X' = P(t)X$, $X(0) = I$, where $P(t)$ is periodic of period one, may be written in the form $X(t) = Q(t)e^{Ct}$, where $Q(t)$ is periodic of period one.

37. The determinant

$$W(t) = W(u_1, u_2, \ldots, u_N) = \begin{vmatrix} u_1 & u_2 & \cdots & u_N \\ u_1' & u_2' & \cdots & u_N' \\ \vdots & & & \\ u_1^{(N-1)} & u_2^{(N-1)} & \cdots & u_N^{(N-1)} \end{vmatrix}$$

is called the Wronskian of $u_1, u_2, \ldots, u_N$. If the u_i are solutions of $u^{(N)} + p_1 u^{(N-1)} + \cdots + p_N u = 0$, show that

$$W(t) = W(0) \exp\left(-\int_0^t p_1(s) \, ds\right).$$

BIBLIOGRAPHY AND COMMENTS

For more detailed results, see

R. Bellman, *Introduction to Matrix Analysis*, McGraw-Hill, New York, 1960.

6.18. For an account of the interconnections between the matrix Riccati equation and two-point boundary value problems, see

W. T. Reid, "Solution of a Riccati Matrix Differential Equation as Functions of Initial Values," *J. Math. Mech.*, **8**, 1959, pp. 221–230.

W. T. Reid, "Properties of Solutions of a Riccati Matrix Differential Equation," *J. Math. Mech.*, **9**, 1960, pp. 749–770.

W. T. Reid, "Oscillation Criteria for Self-Adjoint Differential Systems," *Trans. Amer. Math. Soc.*, **101**, 1961, pp. 91–106.

W. T. Reid, "Principal Solutions of Nonoscillatory Linear Differential Systems," *J. Math. Anal. Appl.*, **9**, 1964, pp. 397–423.

W. T. Reid, "Riccati Matrix Differential Equations and Nonoscillation Criteria for Associated Linear Differential Systems," *Pacific J. Math.*, **13**, 1963, pp. 665–685.

W. T. Reid, "A Class of Two-Point Boundary Problems," *Illinois J. Math.*, **2**, 1958, pp. 434–453.

W. T. Reid, "Principal Solutions of Nonoscillatory Self-Adjoint Linear Differential Systems," *Pacific J. Math.*, **8**, 1958, pp. 147–169.

W. T. Reid, "A Class of Monotone Riccati Matrix Differential Operators," *Duke Math. J.*, **32**, 1965, pp. 689–696.

See also

M. Morse, *The Calculus of Variations in the Large* (Colloq. Publ., Vol. 18), Amer. Math. Soc., Providence, Rhode Island, 1960.

E. C. Tomastik, "Singular Quadratic Functionals of n Dependent Variables," *Trans. Amer. Math. Soc.*, **124**, 1966, pp. 60–76.

Our aim throughout has not been to handle the most general problem, but rather to provide a sufficiently firm basis so that the reader can penetrate into newer and more complex areas with a reasonable degree of assurance.

7

MULTIDIMENSIONAL CONTROL PROCESSES VIA

THE CALCULUS OF VARIATIONS

7.1. Introduction

As pointed out in the introduction, a basic problem in control theory is that of minimizing the functional

$$J(x_1, x_2, \ldots, x_N; y_1, y_2, \ldots, y_N)$$
$$= \int_0^T g(x_1, x_2, \ldots, x_N; y_1, y_2, \ldots, y_N; t) \, dt \qquad (7.1.1)$$

over all admissible functions y_i, where the x_i and the y_i are connected by the system of differential equations

$$\frac{dx_i}{dt} = h_i(x_1, x_2, \ldots, x_N; y_1, y_2, \ldots, y_N; t), \quad x_i(0) = c_i. \qquad (7.1.2)$$

Introducing vector notation, we can write

$$J(x, y) = \int_0^T g(x, y, t) \, dt \qquad (7.1.3)$$

and

$$\frac{dx}{dt} = h(x, y, t), \qquad x(0) = c. \qquad (7.1.4)$$

This variational problem, oversimplified as it is as far as describing realistic control processes, is nonetheless a formidable mathematical problem, both analytically and computationally. Consequently, we begin with a simpler version that, as pointed out previously, can be used to resolve the original problem by means of successive approximations in carefully chosen situations. The demonstration of this will be given in Volume II.

Let us then study the problem of minimizing

$$J(x, y) = \int_0^T [(A(t)x, x) + (y, y)] \, dt, \qquad (7.1.5)$$

where

$$x' = Bx + y, \qquad x(0) = c. \qquad (7.1.6)$$

We shall discuss first the case where $B = 0$ and $A(t)$ is a constant matrix, then the case where $B = 0$ and $A(t)$ is variable, and then the case where $B \neq 0$. In dealing with matrix equations, there is a considerable point to treating various special cases before boldly challenging the general case.

At the end of the chapter, we will discuss minimization subject to constraints using the Lagrange multiplier and using the theory of inequalities.

A number of results discussed in detail in Chapter 4 will be relegated to the exercises here.

7.2. The Euler Equation

Consider the functional

$$J(x) = \int_0^T [(\dot{x}, \dot{x}) + (A(t)x, x)] \, dt, \qquad (7.2.1)$$

corresponding to the special case $B = 0$ of (7.1.6). The admissible functions are now those for which $\int_0^T (\dot{x}, \dot{x}) \, dt$ exists as a Lebesgue integral. This imposes no requirement of continuity on $\dot{x}$. Suppose that $\bar{x}$

minimizes $J(x)$, subject to $x(0) = c$, and set $x = \bar{x} + \varepsilon y$, where ε is a scalar and y is an arbitrary vector function of t. Then

$$J(\bar{x} + \varepsilon y) = J(\bar{x}) + \varepsilon^2 J(y) + 2\varepsilon \int_0^T [(\dot{\bar{x}}, \dot{y}) + (A(t)\bar{x}, y)] \, dt. \tag{7.2.2}$$

If $J(\bar{x})$ is to be a minimum, we must have

$$\int_0^T [(\dot{\bar{x}}, \dot{y}) + (A(t)\bar{x}, y)] \, dt = 0. \tag{7.2.3}$$

Integrating by parts, this becomes

$$(\dot{\bar{x}}, \dot{y})]_0^T + \int_0^T [-(\ddot{\bar{x}}, y) + (A(t)\bar{x}, y)] \, dt$$

$$= (\dot{\bar{x}}, \dot{y})]_0^T + \int_0^T [(-\ddot{\bar{x}} + A(t)x, y)] \, dt = 0. \tag{7.2.4}$$

Since this holds for all y, we expect that the conditions on $\bar{x}$ are

$$x'' - A(t)x = 0, \qquad x(0) = c, \qquad x'(T) = 0. \tag{7.2.5}$$

We can now safely drop the bar. This is the Euler equation. As before, it is the equation we want and we are not concerned with the rigorous aspects of the derivation.

EXERCISES

1. Consider the minimization of $J(u) = \int_0^T (u''^2 + u^2) \, dt$, where $u(0) = c_1, u'(0) = c_2$. Obtain the Euler equation and the asymptotic behavior of the minimum value of $J(u)$ as $T \to \infty$.
2. Consider the minimum value of $J(u)$ subject to $u(0) = c_1, u'(0) = c_2$, $u''(0) = c_3$. Show that the constraint $u''(0) = c_3$ may be discarded.
3. Use the Haar device to obtain (7.2.5) rigorously.

7.3. The Case of Constant A

Let us suppose that $A(t)$ is constant, and that it is a positive definite matrix. The Euler equation is then

$$x'' - Ax = 0, \qquad x(0) = c, \qquad x'(T) = 0. \tag{7.3.1}$$

Let $S = A^{1/2}$ denote the positive definite square root of A. Then the general solution of the matrix equation

$$X'' - AX = 0 \tag{7.3.2}$$

is given by

$$X = e^{St}C_1 + e^{-St}C_2, \tag{7.3.3}$$

where C_1 and C_2 are constant matrices. The general solution of $x'' - Ax = 0$ is

$$x = e^{St}c^{(1)} + e^{-St}c^{(2)}, \tag{7.3.4}$$

where $c^{(1)}$ and $c^{(2)}$ are constant vectors. Using the boundary conditions of (7.3.1), we have

$$c = c^{(1)} + c^{(2)},$$
$$0 = Se^{ST}c^{(1)} - Se^{-ST}c^{(2)}. \tag{7.3.5}$$

This system of simultaneous equations must be solved for $c^{(1)}$ and $c^{(2)}$.

It is a bit easier if we begin with the general solution of the vector differential equation in the form

$$x = \left(\frac{e^{St} + e^{-St}}{2}\right)b^{(1)} + \left(\frac{e^{St} - e^{-St}}{2}\right)b^{(2)}. \tag{7.3.6}$$

Then the boundary conditions yield

$$c = b^{(1)},$$
$$0 = S\left(\frac{e^{ST} - e^{-ST}}{2}\right)b^{(1)} + S\left(\frac{e^{ST} + e^{-ST}}{2}\right)b^{(2)}. \tag{7.3.7}$$

Since S is positive definite, S^{-1} exists. Therefore, these equations yield

$$0 = \left(\frac{e^{ST} - e^{-ST}}{2}\right)c + \left(\frac{e^{ST} + e^{-ST}}{2}\right)b^{(2)}. \tag{7.3.8}$$

Hence, $b^{(2)}$ is uniquely determined by the expression

$$b^{(2)} = -\left(\frac{e^{ST} + e^{-ST}}{2}\right)^{-1}\left(\frac{e^{ST} - e^{-ST}}{2}\right)c, \tag{7.3.9}$$

provided that the matrix $((e^{ST} + e^{-ST})/2)$ is nonsingular. This we have already noted in the exercises, but we will provide a proof in Section 7.4 below for the sake of completeness.

Let us introduce the matrix functions

$$\sinh X = \frac{e^X - e^{-X}}{2},$$

$$\cosh X = \frac{e^X + e^{-X}}{2}.$$

(7.3.10)

Then (7.3.9) may be written

$$b^{(2)} = -(\cosh XT)^{-1}(\sinh XT)c$$
$$= -(\tanh XT)c.$$

(7.3.11)

The solution of (7.3.1) takes the form

$$x = (\cosh St)c - (\sinh St)(\tanh XT)c$$
$$= (\cosh ST)^{-1}(\cosh S(t - T))c,$$

(7.3.12)

completely analogous to the result obtained in the scalar case. Note that $x'(0) = -S(\tanh ST)c$.

7.4. Nonsingularity of cosh ST

It remains to show that cosh ST is never singular. This follows immediately from the observation that its characteristic roots are $\cosh s_i T$, $i = 1, 2, \ldots, N$, where the s_i are the characteristic roots of S, namely $s_i = \lambda_i^{1/2}$, where the λ_i are the characteristic roots of A. Hence,

$$|\cosh ST| = \prod_{i=1}^{N} \cosh s_i T > 0.$$

(7.4.1)

7.5. The Minimum Value

Starting with

$$x'' - Ax = 0, \qquad x(0) = c, \qquad x'(T) = 0,$$

(7.5.1)

we obtain the relation

$$\int_0^T (x, (x'' - Ax))\, dt = 0, \qquad (7.5.2)$$

and thus

$$(x, x')]_0^T - \int_0^T [(x', x') + (x, Ax)]\, dt = 0. \qquad (7.5.3)$$

Thus,

$$\min_x J(x) = -(c, x'(0))$$

$$= (c, S(\tanh ST)c). \qquad (7.5.4)$$

As before, the relation in (7.2.2) shows that this is the absolute minimum under the assumption that A is positive definite. We have

$$J(x + y) = J(x) + J(y) + 2 \int_0^T [(\dot{x}, \dot{y}) + (x, Ay)]\, dt. \qquad (7.5.5)$$

Integrating by parts, we see that the third term vanishes.

7.6. Asymptotic Behavior

As $T \to \infty$, we have

$$\tanh ST = (e^{ST} + e^{-ST})^{-1}(e^{ST} - e^{-ST})$$

$$\cong I. \qquad (7.6.1)$$

Hence,

$$\min_x J(x) \cong (c, Sc) \qquad (7.6.2)$$

as $T \to \infty$, where $S = A^{1/2}$ as above, and

$$x'(t) \cong -Sx(t). \qquad (7.6.3)$$

Once again, we observe that the asymptotic control law is one of great simplicity.

7.7. Variable $A(t)$

Let us now consider the equation of (7.2.5) in the case where $A(t)$ is variable. Let X_1, X_2, respectively, denote the principal solutions of

$$X'' - A(t)X = 0, \qquad (7.7.1)$$

that is to say, those solutions satisfying the initial conditions

$$
\begin{aligned}
X_1(0) &= 1, & X_1'(0) &= 0, \\
X_2(0) &= 0, & X_2'(0) &= 1.
\end{aligned}
\qquad (7.7.2)
$$

Then the general solution of

$$x'' - A(t)x = 0 \qquad (7.7.3)$$

is given by

$$x = X_1(t)c^{(1)} + X_2(t)c^{(2)}. \qquad (7.7.4)$$

The boundary conditions $x(0) = c$, $x'(T) = 0$ yield the relations

$$
\begin{aligned}
c &= c^{(1)}, \\
0 &= X_1'(T)c^{(1)} + X_2'(T)c^{(2)},
\end{aligned}
\qquad (7.7.5)
$$

whence

$$c^{(2)} = -X_2'(T)^{-1}X_1'(T)c, \qquad (7.7.6)$$

provided that $X_2'(T)$ is nonsingular. Thus,

$$x = [X_1(t) - X_2(t)X_2'(T)^{-1}X_1'(T)]c, \qquad (7.7.7)$$

and $x'(0)$, the missing initial value, is given by

$$x'(0) = -X_2'(T)^{-1}X_1'(T)c. \qquad (7.7.8)$$

7.8. The Nonsingularity of $X_2'(T)$

To show that $X_2'(T)$ is nonsingular, we generalize an appropriate argument used in the scalar case. If $|X_2'(T)| = 0$, there is a nontrivial vector b such that $X_2'(T)b = 0$. Let

$$y = X_2(t)b. \qquad (7.8.1)$$

Then $y(0) = 0$, $y'(T) = 0$, and $y'' - A(t)y = 0$. Thus

$$\int_0^T (y, y'' - A(t)y) \, dt = 0 = (y, y')]_0^T - \int_0^T [(y', y') + (y, A(t)y)] \, dt.$$

$$(7.8.2)$$

Hence, if we assume that

$$J(y) = \int_0^T [(y', y') + (y, A(t)y)] \, dt \qquad (7.8.3)$$

is positive for all nontrivial y satisfying the conditions $y(0) = y'(T) = 0$, we have a contradiction.

EXERCISE

1. Suppose we proceed in the following way. If the minimizing function satisfies the equation $x'' - A(t)x = 0$, $x(0) = c$, it must have the form $x = X_1(t)c + X_2(t)a$, where a is to be determined. Write $J(x) = J(X_1(t)c + X_2(t)a)$ and minimize over the vector a. This determines a uniquely. Obtain the minimizing value of a and show in this fashion that $X_2'(T) \neq 0$.

7.9. The Minimum Value

We have

$$\int_0^T (x, x'' - A(t)x) \, dt = 0 = (x, x')]_0^T - \int_0^T [(x', x') + (x, A(t)x)] \, dt.$$

$$(7.9.1)$$

Hence,

$$\min_x J(x) = -(c, x'(0))$$
$$= (c, X_2'(T)^{-1}X_1'(T)c)$$
$$= (c, R(T)c), \qquad (7.9.2)$$

where $R(T) = X_2'(T)^{-1}X_1'(T)$.

EXERCISE

1. Show by direct calculation that $R(T)$ satisfies a matrix Riccati equation as a function of T.

7.10. Computational Aspects

One numerical resolution of the problem depends upon the determination of the solution of the Euler equation

$$x'' - A(t)x = 0 \tag{7.10.1}$$

subject to the initial conditions $x(0) = c$, $x'(0) = X_2'(T)^{-1}X_1'(T)c$.

There are several aspects to a numerical solution following this route. There is the calculation of $X_1'(T)$ and $X_2'(T)$, the calculation of $X_2'(T)^{-1}$, and finally the determination of $x(t)$ using (7.10.1).

The evaluation of $X_1'(T)$ and $X_2'(T)$ can be accomplished by means of the simultaneous solution of $2N^2$ linear differential equations of the first order,

$$\begin{aligned} X' &= Y, & X(0) &= 0, \\ Y' &= A(t)X, & Y(0) &= I, \end{aligned} \tag{7.10.2}$$

or we can solve N sets of $2N$ simultaneous linear differential equations. These are the equations for the columns of X and Y.

With digital computers currently available, this is pretty much of a routine exercise for N of the order of 100 to 1000. In the near future, it will be equally routine for N of the order 10,000 to 100,000.

Difficulties, however, can easily arise if T is large. To see this, consider the case where $A(t)$ is constant and equal to a matrix A that is positive definite. As $T \to \infty$, the minimizing function behaves like $e^{-St}c/2$, for t not close to the terminal value T. Here, $S = A^{1/2}$. Inevitably, however, in the course of the calculation, numerical errors will arise, introducing an undesired $e^{St}\varepsilon$, where ε is a vector of small norm. If t is large, this term dominates the solution. In other words, as discussed in Chapter 4, the Euler equation is numerically unstable.

There are various ways of countering this characteristic of many of the two-point boundary problems of mathematical physics. The simplest is to avoid it by means of either dynamic programming or invariant imbedding. We will pursue this route in Chapter 8.

If A is constant, we can avoid the numerical integration by using the canonical representation of A in terms of characteristic roots and vectors. In general, numerical integration is required.

The problem of numerical determination of $x(t)$ is compounded by the fact that $x'(0)$ is determined by the expression $-X_2'(T)^{-1}X_1(T)$. There is serious danger of considerable error in the inversion of the $N \times N$ matrix $X_2'(T)$. This is particularly the case since $X_2'(T)$ is practically singular for large T.

The reason for this is the following. The matrix e^{St} has the form

$$e^{St} = T \begin{pmatrix} e^{b_1 t} & & & & \\ & e^{b_2 t} & & & 0 \\ & & \cdot & & \\ & & & \cdot & \\ 0 & & & & \cdot \\ & & & & e^{b_N t} \end{pmatrix} T'. \qquad (7.10.3)$$

Let $0 < b_1 < b_2 < \cdots < b_N$. Then as $t \to \infty$,

$$e^{St} = e^{b_1 t}\gamma_1 + e^{b_2 t}\gamma_2 + \cdots + e^{b_N t}\gamma_N \cong e^{b_N t}\gamma_N, \qquad (7.10.4)$$

where

$$\gamma_N = T \begin{pmatrix} 1 & & & 0 \\ & 0 & & \\ & & \cdot & \\ & & & \cdot \\ 0 & & & \cdot \\ & & & & 0 \end{pmatrix} T', \qquad (7.10.5)$$

a very singular matrix. Thus, e^{ST} can be highly ill-conditioned for large T.

It cannot be overemphasized that no solution of the original variational problem can be called complete unless it leads to feasible computational algorithms.

7.11. $\min_y \int_0^T [(x, x) + (y, y)]\, dt, \quad x' = Bx + y, \quad x(0) = c$

It is quite straightforward to obtain the minimum of

$$J(x, y) = \int_0^T [(x, x) + (y, y)]\, dt \qquad (7.11.1)$$

where x and y are connected by

$$x' = Bx + y, \qquad x(0) = c. \tag{7.11.2}$$

If $\bar{x}, \bar{y}$ are the candidates for an extremal pair, we set $x = \bar{x} + \varepsilon w$, $y = \bar{y} + \varepsilon z$ and obtain in the usual fashion the equation

$$\int_0^T [(x, w) + (y, z)] \, dt = 0, \qquad w' = Bw + z, \qquad w(0) = 0. \tag{7.11.3}$$

Substituting the value $z = w' - Bw$ in the integral relation, we have

$$\int_0^T [(\bar{x}, w) + (\bar{y}, w' - Bw)] \, dt = 0, \tag{7.11.4}$$

whence an integration by parts yields the adjoint equation

$$\bar{y}' = -B'\bar{y} + \bar{x}, \qquad \bar{y}(T) = 0, \tag{7.11.5}$$

where B' denotes here the transpose matrix.

The variational equations are thus, dropping the bars,

$$\begin{aligned} x' &= Bx + y, & x(0) &= c, \\ y' &= -B'y + x, & y(T) &= 0. \end{aligned} \tag{7.11.6}$$

We leave for the reader as a series of exercises the task of demonstrating the existence and uniqueness of solution of this system and the fact that this solution yields the minimum of $J(x, y)$.

Miscellaneous Exercises

1. In a similar fashion carry out the minimization of

$$J(x, y) = \int_0^T [(x, Ax) + (y, By)] \, dt$$

subject to $x' = Cx + Dy$, $x(0) = c$.

2. Carry out in a similar fashion the minimization of

$$J(x, y) = \int_0^T [(x, A(t)x) + (y, B(t)y)] \, dt$$

subject to $x' = C(t)x + D(t)y$, $x(0) = c$.

3. Show how to reduce the foregoing problems in the cases where D and $D(t)$ are nonsingular.

4. Show that if D is singular, the problem is equivalent to one in which y is replaced by a control vector of lower dimension.

5. Consider, in particular, the problem of minimizing

$$\int_0^T \left[\sum_{k=0}^{N-1} (u^{(k)})^2 + v^2 \right] dt,$$

where u is the solution of $u^{(N)} + a_1 u^{(N-1)} + \cdots + a_N u = v$, $u^{(k)}(0) = c_k$, $k = 0, 1, \ldots, N-1$. Obtain the differential equation satisfied by the v that minimizes.

6. Using the representation

$$(c, A^{1/2}c) = \min_x \int_0^\infty ((x', x') + (x, Ax)) \, dt,$$

over $x(0) = c$, conclude that $A > B$ implies $A^{1/2} > B^{1/2}$.

7. Consider the vector equation $x'' - A(t)x = 0$, where $A(t)$ is positive definite for $t \geq 0$. Show that if $x(0) = 0$, there is no point $t_1 > 0$ such that $x(t_1) = 0$, nor one for which $x'(t_1) = 0$. Establish the same results for the case where $x'(0) = 0$. (*Hint:* Consider the integral $\int_0^T (x, x'' - A(t)x) \, dt$.)

8. Consider the problem of minimizing

$$J(x) = \int_0^T [(x - h(t), x - h(t)) + (x', x')] \, dt,$$

a smoothing problem.

9. Consider the problem of minimizing

$$\int_0^T [(x', x') + (x, x) + \varepsilon h(x)] \, dt,$$

where ε is a small parameter and $x(0) = c$. Show that the associated Euler equation $x'' - x - \varepsilon g(x) = 0$, $x(0) = c$, $x'(T) = 0$, has a solution by considering the Jacobian of $x'(T)$ at $t = T$ for $\varepsilon = 0$.

10. What is the behavior of

$$\min_x \left[\int_0^T [\varepsilon(x', x') + (x, Ax)] \, dt \right]$$

as $\varepsilon \to 0$, where $x(0) = c$ and A is positive definite?

11. Consider the problem of minimizing

$$J(x, y) = \int_0^T [(x, x) + (y, y)] \, dt$$

with respect to y, where $x' = Ax + y$, and $x(0) = c_1$, $x(T) = c_2$. To avoid the problem of determining where y can be chosen so as to meet the second condition, $x(T) = c_2$, consider the problem of minimizing

$$J(x, y, \lambda) = \int_0^T [(x, x) + (y, y)] \, dt + \lambda(x(T) - c_2, x(T) - c_2)$$

for $\lambda \geq 0$, where the only constraint is now $x(0) = c_1$. Study the asymptotic behavior of $\min_y J(x, y, \lambda)$ as $\lambda \to \infty$ and thus obtain a sufficient condition that a control y exist such that $x(T) = c_2$.

(This is connected with a property of a system called "controllability" which has been intensively studied over the past few years.)

8

MULTIDIMENSIONAL CONTROL PROCESSES VIA

DYNAMIC PROGRAMMING

8.1. Introduction

In this chapter, we wish to discuss multidimensional control processes using the theory of dynamic programming. We shall begin with a consideration of the minimization of the scalar functional

$$J(x, y) = \int_0^T [(x, Ax) + (y, y)] \, dt, \qquad (8.1.1)$$

where x and y are connected by the differential equation

$$x' = Bx + y, \qquad x(0) = c. \qquad (8.1.2)$$

As before, we will rely upon our knowledge of the analytic structure of the minimum value of $J(x, y)$ to make our formal procedures rigorous. This is the easiest route to pursue, but not the only one.

Having obtained a multidimensional Riccati equation, following a familiar path, we next examine the question of the feasibility of a numerical solution of this equation. A particular aspect of this study of some importance is that of determining the asymptotic control law. This leads us into questions of extrapolation and effective extension of the radius of convergence of power series.

Following this, we shall discuss the minimization of

$$J(x) = \int_0^T [(x', x') + (x, Ax)] dt \qquad (8.1.3)$$

subject to integral constraints such as

$$\int_0^T (x, f_i(t_1)) \, dt_1 = a_i, \qquad i = 1, 2, \ldots, R, \qquad (8.1.4)$$

using the theory of inequalities, as in Chapter 4.

Finally, we will turn our attention to discrete control processes, and indicate some applications of this theory to the solution of ill-conditioned systems of linear algebraic equations.

8.2. $\int_0^T [(x', x') + (x, Ax)] dt$

Let us begin with the case where the matrix B in (8.1.2) is taken to be zero. Then the problem is that of minimizing

$$J(x) = \int_0^T [(x', x') + (x, Ax)] dt \qquad (8.2.1)$$

where A is taken to be positive definite.

Write

$$f(c, T) = \min_x J(x). \qquad (8.2.2)$$

Then, using the symbolic diagram of Fig. 8.1, where $v = v(c, T)$ represents the value of $x'(0)$, the initial "slope," the principle of optimality yields the relation

$$f(c, T) = \min_v [\{(v, v) + (c, Ac)\}\Delta + f(c + v\Delta, T - \Delta)], \qquad (8.2.3)$$

to terms that are order of Δ^2.

The multidimensional Taylor expansion yields

$$f(c + v\Delta, T - \Delta) = f(c, T) - \Delta \frac{\partial f}{\partial T} + \Delta(v, \mathrm{grad}\, f) + O(\Delta^2). \qquad (8.2.4)$$

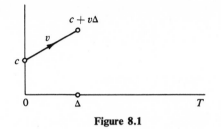

Figure 8.1

Hence, cancelling the common term $f(c, T)$ and letting $\Delta \to 0$, we obtain the nonlinear partial differential equation

$$\frac{\partial f}{\partial T} = \min_{v} \left[(v, v) + (c, Ac) + (v, \text{grad } f) \right]. \qquad (8.2.5)$$

All of this is completely parallel to what we have done in the scalar case.

The minimization over v is easily effected. The minimum is assumed at

$$v = -\frac{\text{grad } f}{2}, \qquad (8.2.6)$$

yielding the quadratically nonlinear partial differential equation

$$\frac{\partial f}{\partial T} = (c, Ac) - \frac{(\text{grad } f, \text{grad } f)}{4}. \qquad (8.2.7)$$

The initial condition is readily seen to be $f(c, 0) = 0$.

8.3. The Associated Riccati Equation

Fortunately, we know that

$$f(c, T) = (c, R(T)c), \qquad (8.3.1)$$

where $R(T)$ is a positive definite matrix. Then

$$\text{grad } f = 2R(T)c, \qquad (8.3.2)$$

and (8.2.7) yields the ordinary differential equation

$$R' = A - R^2, \qquad R(0) = 0, \qquad (8.3.3)$$

a matrix Riccati equation. The policy assumes the simple form

$$v(c, T) = -R(T)c, \qquad (8.3.4)$$

a linear control policy.

EXERCISES

1. Show that (8.3.3) can be solved in terms of the solutions of $X'' - AX = 0$.
2. Consider the problem of minimizing $J(u) = \int_0^T (u''^2 + u^2)\, dt$, where $u(0) = c_1, u'(0) = c_2$. What is the vector-matrix version?

8.4. Asymptotic Behavior

From the differential equation of (8.3.3), we see that if $R(T)$ approaches a limit as $T \to \infty$, this limit must be $A^{1/2}$, the positive definite square root of A. Since $R(T)$ is monotone increasing and uniformly bounded (as we see by taking $x = e^{-t}c$), the limit exists. To determine the rate of approach, set

$$R = A^{1/2} - S, \qquad (8.4.1)$$

where $S(T)$ is a new variable. Then

$$-\frac{dS}{dT} = A - (A^{1/2} - S)^2$$

$$= A - A + A^{1/2}S + SA^{1/2} - S^2. \qquad (8.4.2)$$

Hence, neglecting the second-order term S^2, we have the approximate linear differential equation

$$\frac{dS_1}{dT} = -A^{1/2}S_1 - S_1 A^{1/2}. \qquad (8.4.3)$$

The solution of this is, as we know,

$$S_1 = \exp(-A^{1/2}t)S_1(0)\exp(-A^{1/2}T). \qquad (8.4.4)$$

This shows that $\|S_1\| = O[\exp(-2\lambda_1^{1/2}T)$ as $T \to \infty$, where λ_1 is the smallest characteristic root of A.

Since the optimal policy has the form

$$v(c, T) = -R(T)c, \tag{8.4.5}$$

we see that as $T \to \infty$,

$$v(c, T) = -R(\infty)c + O[\exp(-2\lambda_1^{1/2}T)]$$
$$= -A^{1/2}c + O[\exp(-2\lambda_1^{1/2}T)]. \tag{8.4.6}$$

It follows that the steady-state policy is an excellent approximation for large T.

8.5. Rigorous Aspects

The equation $R(T)$ was derived in a formal fashion in Section 8.2. We can pursue two different paths at this juncture, to establish the equation rigorously. Either we can use our knowledge of the explicit value of $R(T)$, namely, $A^{1/2} \tanh A^{1/2}T$, to show by direct calculation that (8.3.3) is satisfied or we can use the fact that $f(c, T) = (c, R(T)c)$, and that an optimal policy exists, to rigorize the argument in Section 8.2.

It is important to point out, of course, that there is a considerable difference between validating a result once obtained and a method that generates the results in a routine fashion.

8.6. Time-Dependent Case

Let us now consider the case where A is dependent on t. Write

$$f(a, c) = \min_x \int_a^T [(x', x') + (x, A(t)x)] \, dt, \tag{8.6.1}$$

subject to the initial condition $x(a) = c$. The function $f(a, c)$ is defined for all real c and for $0 \le a \le T$. Proceeding as before, we have

$$f_a = -\min_v [(v, v) + (c, A(a)c) + (v, \text{grad} f)]$$
$$= -(c, A(a)c) + \frac{(\text{grad} f, \text{grad} f)}{4}, \tag{8.6.2}$$

with the initial condition $f(T, c) = 0$. Using the quadratic structure $f(a, c) = (c, R(a)c)$, we obtain the ordinary differential equation

$$R'(a) = -A(a) + R(a)^2, \qquad R(T) = 0. \tag{8.6.3}$$

EXERCISES

1. Show that (8.6.3) can be solved in terms of the solution of $X'' - A(t)X = 0$; show that $R(a) = Y_1'(a)Y_1(a)^{-1}c$, where $Y_1(t)$ is the principal solution of the matrix solution starting from T.
2. Show that

$$f(c, T) = \min_y \int_0^T [(x, Ax) + (y, y)] \, dt,$$

where $x' = Bx + y$, $x(0) = c$, satisfies the equation $f_T = (c, Ac) + (Bc, \text{grad } f) - (\text{grad } f, \text{grad } f)/4$, $f(c, 0) = 0$. Setting $f(c, T) = (c, R(T)c)$, show that $R' = A + 2BR + 2RB' - R^2$, $R(0) = 0$.
3. Obtain the value of $R(\infty)$ by setting $R = 2B + S$.
4. Show that changes of variable in both x and y convert the problem of minimizing $\int_0^T [(x, Ax) + (y, By)] \, dt$ subject to $x' = Cx + Dy$, $x(0) = c$ into the problem treated in the previous section, provided that $B > 0$ and D is nonsingular.
5. Consider what reductions are possible in connection with the minimization of $\int_0^T [(x, A(t)x) + (y, B(t)y)] \, dt$ subject to $x' = C(t)x + D(t)y$, $x(0) = c$.

8.7. Computational Aspects

Let us now examine some of the computational aspects of the determination of $R(T)$. Consider the equation

$$R' = A - R^2, \qquad R(0) = 0. \tag{8.7.1}$$

If R is of dimension N, we are faced with the problem of the numerical integration of N^2 simultaneous ordinary differential equations subject to initial conditions. Taking account of the symmetry of R, we can reduce this to $N(N + 1)/2$. If N is not large it may be preferable to keep the entire set of N^2 equations for the sake of an internal check on the accuracy of the calculations.

If N is large, for example, $N = 100$, we must deal with at least 5050 simultaneous equations. This is a number that commands a certain respect even today. In any case, there is always the problem of minimizing the computing time, particularly in connection with on-line control.

If we employ the approach of Chapter 7, we have the privilege of integrating vector systems of order N, but, in return, eventually we find it necessary to solve an $N \times N$ system of linear algebraic equations. This is always a matter of some delicacy, even when N is small. This last fact is an important point to keep in mind when the control problem treated in this and the preceding chapters is merely one stage of a much larger problem. This is the situation, for example, when the method of successive approximations is employed to treat more general control processes.

Consequently, it is fair to state that high-dimensional control processes, which is to say, realistic control processes, conjure up computational difficulties regardless of the approach used in their treatment. Fortunately, as digital computers become more and more powerful, all approaches become more efficient.

The dynamic programming approach possesses two advantages. The first of these is that of avoiding the solution of linear algebraic equations and the second lies in the inherent stability of the Riccati equation. Principally against the method is the task of integrating a system of N^2, or $N(N + 1)/2$, simultaneous differential equations. Let us see what we can do to circumvent this obstacle to progress.

8.8. Successive Approximations

The advantage in reducing the question to that of solving a linear matrix equation

$$X' = P(t)X, \qquad X(0) = c, \tag{8.8.1}$$

lies in the fact that we can determine X column-by-column, N successive N-dimensional problems rather than one N^2-dimensional problem.

Let us then think in terms of some method of successive approximations that involves linear differential equations. Thus, for example, we can use

$$R_N' = A - R_N R_{N-1}, \qquad R_N(0) = 0, \tag{8.8.2}$$

for $N \geq 1$, with $R_0(T)$ chosen in a judicious fashion. Problems of storage of R_{N-1} while computing R_N arise in this type of procedure.

If we wish to accelerate the convergence, we can employ quasi-linearization,

$$R_N' = A + R_{N-1}^2 - R_N R_{N-1} - R_{N-1}R_N, \qquad R_N(0) = 0. \quad (8.8.3)$$

As we shall see in the next section, this equation has an interesting interpretation.

8.9. Approximation in Policy Space

In place of approximation of a solution in the space of the return matrices $R(T)$, let us think in terms of an approximation in the space of control policies. Let us then return to the equation

$$f_T = \min_v \; [(c, c) + (v, v) + (v, \operatorname{grad} f)] \quad (8.9.1)$$

and investigate the possibility of choosing a sequence of control functions $v(c, T)$. Writing $f(c, T) = (c, R(T)c)$, (8.9.1) becomes

$$(c, R'(T)c) = \min_v \; [(c, c) + (v, v) + 2(v, R(T)c)]. \quad (8.9.2)$$

Suppose that we use an initial guess $v = -S(T)c$, yielding a function $f_1(c, T) = (c, R_1(T)c)$. Then (8.9.2) yields

$$R_1' = I + S^2 - SR_1 - R_1 S, \qquad R_1(0) = 0. \quad (8.9.3)$$

Suppose further that we take as our next estimate an optimal control policy of the form

$$v = -R_1(T)c. \quad (8.9.4)$$

Then (8.9.2) yields

$$R_2' = I + R_1^2 - R_2 R_1 - R_1 R_2, \qquad R_2(0) = 0. \quad (8.9.5)$$

Continuing in this fashion, we obtain the recurrence relation of (8.8.3). We see then that this equation corresponds to a particular type of approximation in policy space.

EXERCISE

1. Is there any advantage to using $R_1, R_2, \ldots, R_N$ to estimate the new linear control law $v_{N+1} = -S_{N+1}(T)c$?

8.10. Monotone Convergence

From the general character of approximation in policy space, we suspect that the convergence is monotone; that is,

$$R_1 \geq R_2 \geq \cdots \geq R_N \geq \cdots \geq R. \tag{8.10.1}$$

It is not difficult to establish this useful result.

We begin with the result that

$$S^2 - RS - SR \geq -R^2 \tag{8.10.2}$$

for all S, with equality only if $S = R$ where we assume that R and S are both symmetric matrices. This follows from

$$(S - R)^2 \geq 0, \tag{8.10.3}$$

which is equivalent to (8.10.2) when the multiplication is carried out. Hence,

$$S^2 - R_1 S - SR_1 \geq R_1{}^2 - R_1 R_1 - R_1 R_1. \tag{8.10.4}$$

Consider then the two differential relations

$$\frac{dR_1}{dt} = I + S^2 - R_1 S - SR_1, \qquad R_1(0) = 0,$$

$$\frac{dR_1}{dt} \geq I + R_1{}^2 - R_1 R_1 - R_1 R_1. \tag{8.10.5}$$

It follows that $R_1 \geq R_2$, where R_2 is the solution of the equality

$$\frac{dR_2}{dt} = I + R_1{}^2 - R_2 R_1 - R_1 R_2, \qquad R_2(0) = 0. \tag{8.10.6}$$

The argument concerning the monotonicity of the R_n now proceeds intuitively. Since the R_n are positive definite and monotone decreasing, they converge, necessarily, to $R(T)$. The convergence is thus monotone and quadratic. This last problem is readily established as in the scalar case.

The foregoing shows that we possess a systematic method for improving any initial policy $y = -S(T)c$.

EXERCISE

1. Extend (8.10.2) by considering $(S - R)(S' - R) \geq 0$ for arbitrary S, where S' denotes the transpose of S.

8.11. Partitioning

Let us now consider an alternative approximation scheme for the case where N is large. Consider the problem of minimizing

$$J(x) = \int_0^T [(x', x') + (x, Ax)] \, dt. \tag{8.11.1}$$

Let us write

$$x = \begin{pmatrix} y \\ z \end{pmatrix}, \qquad A = \begin{pmatrix} A_1 & A_2 \\ A_2 & A_3 \end{pmatrix}, \tag{8.11.2}$$

where the dimensions of y, z, A_1, A_2, and A_3 are $N/2$. Here

$$y = \begin{pmatrix} x_1 \\ x_2 \\ \vdots \\ x_{N/2} \end{pmatrix}, \qquad z = \begin{pmatrix} x_{N/2+1} \\ \vdots \\ x_N \end{pmatrix}. \tag{8.11.3}$$

Then, using inner products in the lower-dimensional spaces, we have

$$J(x) = J(y, z) = \int_0^T [(y', y') + 2(y', z') + (z', z') + (y, A_1 y)$$

$$+ 2(y, A_2 z) + (z, A_3 z)] \, dt. \tag{8.11.4}$$

Let us now guess an initial approximation y_0 and consider the problem of minimizing $J(y_0, z)$ over z. The dimension of the Riccati equation that arises in this connection is $N^2/4$, a considerable reduction if N is large.

The foregoing determines a vector z_0. Next we minimize $J(y, z_0)$, which determines y_1, and so forth. Clearly,

$$J(y_0, z_0) \geq J(y_1, z_0) \geq J(y_1, z_1) \geq \cdots. \tag{8.11.5}$$

If N is quite large, division of x into more than two parts may be necessary, or, equivalently, y may be of much smaller dimension than z. Furthermore, we may pick different subsets of components at each stage depending upon our observation of the changes in the approximations.

Observe once again that we automatically ensure improvement over any existing nonoptimal control policy. In Volume II, we will discuss the question of convergence of this and other approximation procedures and methods of accelerating the convergence.

8.12. Power Series Expansions

Let us now examine the possibility of obtaining the solution of

$$R' = A - R^2, \qquad R(0) = 0, \qquad (8.12.1)$$

by means of a power series expansion†

$$R = B_1 t + B_2 t^2 + \cdots + B_k t^k + \cdots . \qquad (8.12.2)$$

Substituting in (8.12.1), we see that

$$
\begin{aligned}
B_1 &= A, \\
2B_2 &= 0 \\
3B_3 &= - B_1{}^2, \\
4B_4 &= 0, \cdots .
\end{aligned}
\qquad (8.12.3)
$$

What is the radius of convergence? The solution of (8.12.1) is

$$R = A^{1/2} \tanh A^{1/2} t = A^{1/2} T \begin{pmatrix} \tanh \lambda_1^{1/2} t & & \\ & \ddots & \\ & & \tanh \lambda_N^{1/2} t \end{pmatrix} T'. \qquad (8.12.4)$$

If we suppose that $0 < \lambda_1 \le \lambda_2 \le \cdots \le \lambda_N$, we see that the singularity nearest to the origin in the complex t-plane is located at

$$2\lambda_N^{1/2} t = \pm \pi i. \qquad (8.12.5)$$

Hence, the radius of convergence is $\pi/2\lambda_N^{1/2}$.

† We have replaced the time variable T by t in order not to confuse it with the orthogonal matrix T used in the canonical representation of A, $A = TAT'$.

As we shall see in the next section, we may be able to use the values obtained for small t to predict the asymptotic control policy quite accurately.

EXERCISES

1. Let tanh B be the matrix function defined above for a positive definite matrix B. Show that tanh $2B = (2 \tanh B)(I - (\tanh B)^2)^{-1}$.
2. Hence show how to calculate tanh $A^{1/2}t$ for t outside the radius of convergence of the series in (8.12.2).
3. How many matrix inversions are required to go from t_0 to $2^n t_0$?

8.13. Extrapolation

We can find the asymptotic control policy at the expense of determining the characteristic roots and vectors of a symmetric matrix A. Alternatively, we can solve a system of ordinary differential equations,

$$R' = A - R^2, \qquad R(0) = 0. \tag{8.13.1}$$

Both procedures have advantages and disadvantages.

Suppose that the bottleneck is time rather than storage. Can we obtain a good estimate for $R(\infty)$ without the necessity of dealing with characteristic roots and vectors or of integrating (8.13.1) over a long T-interval, an enterprise both time-consuming and risky because of numerical error?

The basic idea is to use an extrapolation technique after having calculated $R(T)$ for small T either by a power series expansion as indicated in the previous section or by means of standard numerical techniques. Effective use of interpolation and extrapolation techniques depends upon extensive knowledge of the analytic structure of the function under study.

In our case we know from the previous analysis that

$$R(T) \sim R(\infty) + R_1 e^{-2\mu_1 T} + \cdots \tag{8.13.2}$$

as $T \to \infty$, where $\mu_1 = \lambda_1^{1/2}$ is the smallest characteristic root of $A^{1/2}$; equivalently, λ_1 is the smallest characteristic root of A. Although the value of $R(\infty)$ is easily seen using (8.13.1), the same is not true of R_1.

There are several paths we can pursue, depending upon whether we wish to calculate μ_1 in advance or not. This may not be an onerous chore since there are several excellent procedures available for calculating the smallest characteristic root of a positive definite matrix.

Assume that we do know μ_1 and let us see how we can estimate $R(\infty)$. If (8.13.2) is actually an equality, we have

$$
\begin{aligned}
R(T) &= R(\infty) + R_1 e^{-2\mu_1 T}, \\
R'(T) &= -2\mu_1 R_1 e^{-2\mu_1 T},
\end{aligned}
\tag{8.13.3}
$$

whence

$$
2\mu_1 R(T) + R'(T) = 2\mu_1 R(\infty).
\tag{8.13.4}
$$

Thus, instead of using the approximation $R(\infty) \cong R(T)$ for large T, we employ (8.13.4). This is a better estimate, since if $\mu_2 = \lambda_2^{1/2}$, we see that

$$
\begin{aligned}
\|R(T) - R(\infty)\| &= O(e^{-2\mu_1 T}), \\
\|2\mu_1 R(T) + R'(T) - R(\infty)\| &= O(e^{-(\mu_1 + \mu_2)T}).
\end{aligned}
\tag{8.13.5}
$$

Similarly, if we calculate μ_2 and write

$$
R(T) = R(\infty) + R_1 e^{-2\mu_1 T} + R_2 e^{-(\mu_1 + \mu_2)T},
\tag{8.13.6}
$$

we can obtain an estimate for $R(\infty)$ in terms of R, R', and R'' that is even more accurate.

If we prefer to avoid the task of determining μ_1, it is convenient to use a nonlinear estimation procedure that is of considerable interest in itself with an extraordinary variety of uses, as the references at the end of the chapter will indicate.

Let us return to the scalar case, which is the same as saying that we will determine $R(\infty)$ element by element. Let $r(T)$ be a typical element. Then the foregoing yields the asymptotic expansion

$$
r(T) \sim r(\infty) + r_1 e^{-2\mu_1 T},
\tag{8.13.7}
$$

which we assume to be an equality,

$$
r(T) = r(\infty) + r_1 e^{-2\mu_1 T},
\tag{8.13.8}
$$

where $r(\infty)$, r_1, and μ_1 are unknown scalars.

Differentiation yields

$$r'(T) = -2\mu_1 r_1 e^{-2\mu_1 T},$$
$$r''(T) = +4\mu_1{}^2 r_1 e^{-2\mu_1 T}. \tag{8.13.9}$$

Here and above, the derivatives are calculated using the differential equation of (8.13.1). Eliminating $r_1 e^{-2\mu_1 T}$ and μ_1, we obtain the expression

$$r(\infty) = \frac{r(T)r''(T) - r'(T)^2}{r''(T)}. \tag{8.13.10}$$

It is easy to verify that

$$\left| r(\infty) - \left(\frac{r(T)r''(T) - r'(T)^2}{r''(T)} \right) \right| = O(e^{-(\mu_1 + \mu_2)T}). \tag{8.13.11}$$

EXERCISES

1. Test the foregoing methods using the scalar equation $r' = 1 - r^2$, $r(0) = 0$.

2. Obtain the nonlinear estimate for $r(\infty)$ based upon the use of

$$r(T) \sim r(\infty) + r_1 e^{-b_1 T} + r_2 e^{-b_2 T}.$$

3. Which is simpler, the use of this estimate or an iteration of the use of (8.3.10)? Which yields a more accurate estimate asymptotically?

4. Derive an analogue of (8.13.10) for the case where we use a discrete set of values $\{r(k\Delta)\}$, $k = 0, 1, 2, \ldots$, to estimate $r(\infty)$.

8.14. Minimization via Inequalities

Let us now consider the problem of minimization of the quadratic functional

$$J(x) = \int_0^T [(x', x') + (x, Ax)] \, dt \tag{8.14.1}$$

over all x satisfying the constraints

$$\int_0^T (x, f_i) \, dt = c_i, \qquad i = 1, 2, \ldots, M. \tag{8.14.2}$$

To illustrate the approach, it is sufficient to consider the case $M = 1$. Introduce the new inner product of two vectors x and y,

$$\langle x, y \rangle = \int_0^T [(x', y') + (x, Ay)] \, dt. \tag{8.14.3}$$

Since $J(x)$ is a positive definite quadratic form in x, by assumption concerning A, we have for any two real scalars r_1 and r_2,

$$J(r_1 x + r_2 y) = r_1{}^2 \langle x, x \rangle + 2r_1 r_2 \langle x, y \rangle + r_2{}^2 \langle y, y \rangle. \tag{8.14.4}$$

Hence, we have the determinantal inequality

$$\begin{vmatrix} \langle x, x \rangle & \langle x, y \rangle \\ \langle x, y \rangle & \langle y, y \rangle \end{vmatrix} \geq 0, \tag{8.14.5}$$

with equality only if $x = r_3 y$ for some scalar r_3.

Since

$$\langle x, y \rangle = (x, y')]_0^T + \int_0^T (x, Ay - y'') \, dt, \tag{8.14.6}$$

we choose y to be the solution of

$$Ay - y'' = f_1, \qquad y'(0) = y'(T) = 0. \tag{8.14.7}$$

We thus obtain the desired minimum value,

$$\min_x J(x) = \frac{c_1{}^2}{\langle y, y \rangle}, \tag{8.14.8}$$

where y is determined by (8.14.7) and the minimizing function $x = r_3 y$, where

$$r_3 \int_0^T (y, f_1) \, dt = c_1. \tag{8.14.9}$$

EXERCISES

1. Minimize $J(x)$ subject to

$$\int_0^T (x, f_1) \, dt = c_1, \qquad \int_0^T (x, f_2) \, dt = c_2,$$

where f_1 and f_2 are linearly independent in $[0, T]$.

2. Using a limiting procedure or otherwise, minimize $J(x)$ subject to

$$\int_0^T (x, f_1)\, dt = c_1, \qquad x(0) = c_2.$$

8.15. Discrete Control Processes

Let us now turn to the study of multidimensional control processes of discrete type. A characteristic problem is that of minimizing the function of $2N$ vector variables

$$J_N(x, y) = \sum_{n=0}^{N} [(x_n, x_n) + (y_n, y_n)] \qquad (8.15.1)$$

over the y_n where

$$x_{n+1} = Ax_n + y_n, \qquad x_0 = c, \qquad (8.15.2)$$

$n = 0, 1, \ldots, N - 1$.

As we shall indicate in Section 8.16 and in subsequent exercises, discrete control processes are important not only in their own right but in relation to other optimization processes that can be interpreted in these terms.

Write

$$f_N(c) = \min_{\{y\}} \sum_{n=0}^{N} [(x_n, x_n) + (y_n, y_n)], \qquad (8.15.3)$$

$N = 0, 1, 2, \ldots$. Then the principle of optimality yields

$$f_N(c) = \min_v [(c, c) + (v, v) + f_{N-1}(Ac + v)], \qquad (8.15.4)$$

$N \geq 1$, with $f_0(c) = (c, c)$. As above, write

$$f_N(c) = (c, R_N c). \qquad (8.15.5)$$

Then

$$(c, R_N c) = \min_v [(c, c) + (v, v) + (Ac + v, R_{N-1}(Ac + v))]. \qquad (8.15.6)$$

The variational equation is

$$v + R_{N-1}(Ac + v) = 0, \qquad (8.15.7)$$

or

$$v = -(I + R_{N-1})^{-1}R_{N-1}Ac. \qquad (8.15.8)$$

Substituting this value back in (8.15.6), we obtain the recurrence relation

$$R_N = (I + A'A) - A'(I + R_{N-1})^{-1}A. \qquad (8.15.9)$$

The algebraic labors are simplified to some extent if we use (8.15.7) before using (8.15.8) in (8.15.6). We have from (8.15.7)

$$(v, v) + (v, R_{N-1}(Ac + v)) = 0,$$
$$(v, v) + (Ac + v, R_{N-1}(Ac + v)) = (Ac, R_{N-1}(Ac + v))$$
$$= -(Ac, v). \qquad (8.15.10)$$

Now employing (8.15.8) we see that

$$-(Ac, v) = (Ac, (I + R_{N-1})^{-1}R_{N-1}Ac). \qquad (8.15.11)$$

Hence

$$(c, R_N c) = (c, c) + (Ac, (I + R_{N-1})^{-1}R_{N-1}Ac), \qquad (8.15.12)$$

and thus

$$R_N = I + A'(I + R_{N-1})^{-1}R_{N-1}A$$
$$= (I + A'A) - A'(I + R_{N-1})^{-1}A. \qquad (8.15.13)$$

It follows that R_N is monotone increasing and uniformly bounded by $I + A'A$. Thus R_N converges as $N \to \infty$. Once again there is an asymptotic control policy, $v \cong -(I + R_\infty)^{-1}R_\infty Ac$, where

$$R_\infty = (I + A'A) - A'(I + R_\infty)^{-1}A. \qquad (8.15.14)$$

EXERCISES

1. Show that R_N can be determined by means of a second-order linear difference equation, and thus determine the rate of convergence of R_N to R_∞.

2. Carry out the same procedure for the more general problem of minimizing

$$J_N(x, y) = \sum_{n=0}^{N} [(x_n, Ax_n) + (y_n, By_n)],$$

where $x_{n+1} = Cx_n + y_n$, $x_0 = c$.

3. Discuss approximation in policy space in connection with the determination of R_∞.

4. Consider the particular case where

$$u_n = u_{n-1} + au_{n-m} + v_n,$$

where $u_0, u_{-1}, \ldots, u_{-m}$ are prescribed and we wish to minimize $J_N(u, v) = \sum_{n=0}^{N} (u_n^2 + v_n^2)$. (See

J. D. R. Kramer, Jr., "On Control of Linear Systems with Time Lags," *Information and Control*, **3**, 1960, pp. 299–326.)

8.16. Ill-Conditioned Linear Systems

As an example of the way in which the analytic machinery of discrete control processes can be employed in other areas, consider the problem of solving the linear system

$$Ax = b. \tag{8.16.1}$$

If A is an ill-conditioned matrix, there are extraordinary obstacles in the path of obtaining an accurate numerical solution. Here "ill-conditioned" is another way of saying "unstable." We mean by this term that small changes in the vector b can produce large changes in x. This is the case if A^{-1} has elements that are large with respect to the order of magnitude of the elements of b and that vary in sign.

A well-known ill-conditioned matrix is $((i + j)^{-1})$, $i, j = 1, 2, \ldots, N$.

One way of tackling the problem of determining x is to consider the associated variational problem of minimizing

$$(Ax - b, Ax - b) + \lambda\varphi(x), \tag{8.16.2}$$

where λ is a scalar and $\varphi(x)$ is chosen to measure some property of the desired x such as smoothness in the dependence of the components on the index, for example,

$$\varphi(x) = (x_2 - x_1)^2 + (x_3 - x_2)^2 + \cdots + (x_N - x_{N-1})^2, \qquad (8.16.3)$$

or closeness to a known approximate solution, for example,

$$\varphi(x) = (x - c, x - c)^2. \qquad (8.16.4)$$

Having achieved this formulation, we now regard the determination of x as a multistage decision process: first x_N, then x_{N-1}, then x_{N-2}, and so on. In the exercises, we will indicate the kinds of results that may be obtained in this fashion.

EXERCISES

1. Consider the problem of minimizing the quadratic form

$$R_N(x) = (Ax - b, Ax - b) + \lambda(x - c, x - c),$$

where we assume that A is nonsingular and $\lambda \geq 0$. Using the explicit form of the vector x, which provides the minimum, $x(\lambda)$, show that $\lim x(\lambda)$ exists as $\lambda \to \infty$ and that this limit is the solution of $Ax = b$.

2. Consider the more general problem of minimizing

$$R_M(x) = [\lambda(x_1 - c_1)^2 + \lambda(x_2 - c_2)^2 + \cdots + \lambda(x_M - c_M)^2$$
$$+ \sum_{i=1}^{N} \left(\sum_{j=1}^{M} a_{ij} x_j - b_i \right)^2],$$

where $M = 1, 2, \ldots, N$. Let $f_M(b) = \min_x R_M(x)$. Then

$$f_1(b) = \min_{x_1} \left[\lambda(x_1 - c_1)^2 + \sum_{i=1}^{N} (a_{i1}x_1 - b_i)^2 \right],$$
$$f_M(b) = \min_{x_M} [\lambda(x_M - c_M)^2 + f_{M-1}(b - x_M a^{(M)})],$$

where $a^{(M)}$ denotes the column vector whose components are $a_{1M}, a_{2M}, \ldots, a_{NM}$.

3. Show that $f_M(b) = (b, Q_M b) + 2(p_M, b) + r_M$ and that

$$Q_M = Q_{M-1} - \frac{A^{(M)}}{(\lambda + K_M)},$$

$$p_M = p_{M-1} - \frac{(\lambda c_M + \rho_M)\alpha^{(M)}}{(\lambda + K_M)},$$

$$r_M = r_{M-1} + \lambda c_M{}^2 - \frac{(\lambda c_M + \rho_M)^2}{(\lambda + K_M)},$$

where

$$\alpha^{(M)} = Q_{M-1}a^{(M)},$$

$$\rho_M = (p_{M-1}, a^{(M)}),$$

$$K_M = (\alpha^{(M)}, a^{(M)}),$$

$$A^{(M)} = \alpha^{(M)} \otimes \alpha^{(M)},$$

where $x \otimes y = (x_i y_j)$, the Kronecker product.

4. Show that the minimizing value of x_M is

$$x_M = \frac{\lambda c_M + (p_{M-1}, a^{(M)}) + (a^{(M)}, Q_{M-1}b)}{\lambda + (a^{(M)}, Q_{M-1}a^{(M)})}.$$

5. Consider the approach to the solution of $Ax = b$ by means of the limit of the sequence $\{x_n\}$, where x_n is determined in terms of x_{n-1} by the requirement that x_n minimize $(Ax - b, Ax - b) + \lambda(x - x_{n-1}, x - x_{n-1})$, with $x_0 = c$. If A is nonsingular and $\lambda \geq 0$, show that the sequence converges regardless of the value of c.

(For application of the foregoing results to the problem of solving an ill-conditioned system $Ax = b$, see

R. Bellman, R. Kalaba, and J. Lockett, "Dynamic Programming and Ill-Conditioned Linear Systems," *J. Math. Anal. Appl.*, **10**, 1965, pp. 206–215.)

6. In place of $R_M(x)$, consider the quadratic form

$$(Ax - y, Ax - y) + \lambda D_N(x),$$

where

$$D_N(x) = (x_1 - x_2)^2 + (x_2 - x_3)^2 + \cdots + (x_{N-1} - x_N)^2.$$

Introduce the sequence of functions $f_k(z, c)$ defined by

$$f_k(z, c) = \min_x \left[\sum_{i=1}^{M} \left(\sum_{j=k}^{N} a_{ij} x_j - z_i \right)^2 \right.$$

$$\left. + \lambda[(x_N - x_{N-1})^2 + \cdots + (x_k - c)^2] \right],$$

$k = 2, 3, \ldots, N$. Then

$$f_N(z, c) = \min_{x_N} \left[\sum_{i=1}^{N} (a_{iN} x_N - z_i)^2 + \lambda(x_N - c)^2 \right],$$

$$f_k(z, c) = \min_{x_k} [\lambda(x_k - c)^2 + f_{k+1}(z - a^{(k)} x_k, x_k)],$$

$k = 2, 3, \ldots, N - 1$, where $a^{(k)}$ has the components $a_{1k}, a_{2k}, \ldots, a_{Mk}$. Write $f_k(z, c) = (z, Q_k z) + 2(z, p_k)c + r_k c^2$ and find recurrence relations for Q_k, p_k, and r_k.

7. Having determined Q_1, p_1, r_1, minimize over c to determine the desired minimization.
(See

R. Bellman, R. Kalaba, and J. Lockett, "Dynamic Programming and Ill-Conditioned Systems—II," *J. Math. Anal. Appl.*, **12**, 1965, pp. 393–400.

See also, for a different approach,

S. Twomey, "On the Numerical Solution of Fredholm Integral Equations of the First Kind by the Inversion of the Linear System Produced by Quadrature," *J. Assoc. Comput. Mach.*, **10**, 1963, pp. 97–101.

References to closely related work by Tychonov will be found there.)

8.17. Lagrange Multipliers

As in the scalar control process, we frequently encounter constraints of the form

$$\int_0^T (x, Fx) \, dt \le k_1, \tag{8.17.1}$$

or, in the discrete case,

$$\sum_{n=0}^{N} (x_n, Fx_n) \le k_2. \tag{8.17.2}$$

One approach to problems of this nature is to use the Lagrange multiplier technique, an obvious extension of the procedure discussed in the scalar case. The proof of monotone behavior of the resource function $g(T) = \int_0^T (x, Fx)\, dt$ can be carried out in much the same manner as in the scalar case. However, a very much simpler proof of the validity of the Lagrange multiplier can be obtained on the basis of the more general ideas of Chapter 9. Consequently, we shall postpone any discussion of the multiplier technique until the following chapter.

8.18. Reduction of Dimensionality

We have discussed in Sections 8.11–8.13 some of the interesting problems connected with obtaining numerical solutions to high-dimensional control processes. Let us now describe a method that can be used in a number of important cases to reduce the dimension of the equation in a drastic fashion and to obtain the solution. As we shall indicate below, it can be considerably augmented by adroit use of the method of successive approximations.

Let the question under discussion be that of minimizing the quadratic functional

$$J(x, y) = (x(T), Bx(T)) + \int_0^T (y, y)\, dt \tag{8.18.1}$$

with respect to y where

$$x' = Ax + y, \qquad x(0) = c. \tag{8.18.2}$$

Proceeding in the usual fashion, we obtain a nonlinear partial differential equation for

$$\min_y J(x, y) = f(c, T) = (c, R(T)c) \tag{8.18.3}$$

which leads to a Riccati differential equation for $R(T)$. If the dimension of x is large, we encounter the difficulties catalogued above in obtaining a numerical solution.

If, however, the matrix B has the form

$$B = \begin{pmatrix} B_k & 0 \\ 0 & 0 \end{pmatrix}, \qquad (8.18.4)$$

where B_k is a $k \times k$ matrix, we can use an approach that leads to a Riccati equation of degree k. If $k \ll N$, we have reduced the problem to manageable terms. Furthermore, we have obtained a more tractable analytic solution.

The criterion function in (8.18.1) is a manifestation of a control process in which the only costs involved are those calculated on the basis of a terminal state at time T and on the basis of the total control exerted over $[0, T]$. In the following section we shall show how the method of successive approximations can be used to treat general control processes of this type in terms of the simple criterion function of (8.18.1), and then to treat quite general control processes in terms of control processes of this restricted type.

To illustrate the method in its simplest form, consider the case where $k = 1$ so that

$$(x(T), Bx(T)) = b_1 x_1(T)^2, \qquad b_1 > 0. \qquad (8.18.5)$$

Solving (8.18.2) for x in terms of y, we have

$$x = e^{At}c + \int_0^t e^{A(t-t_1)}y(t_1)\, dt_1. \qquad (8.18.6)$$

The first component of x, x_1, evaluated at $t = T$, then has the form

$$x_1(T) = (d, c) + \int_0^T (b(t_1), y(t_1))\, dt_1, \qquad (8.18.7)$$

where d is a constant vector and $b(t_1)$ is a variable vector, with the T-dependence suppressed.

Hence,

$$J(x, y) = b_1 \left[(d, c) + \int_0^T (b(t_1), y(t_1))\, dt_1 \right]^2 + \int_0^T (y, y)\, dt, \qquad (8.18.8)$$

with the x-dependence eliminated.

To treat the problem of minimizing this quadratic functional using dynamic programming techniques, we consider the more general problem of minimizing the quadratic functional

$$K(y, u, a) = b_1 \left[u + \int_a^T (b(t_1), y(t_1)) \, dt_1 \right]^2 + \int_a^T (y, y) \, dt, \qquad (8.18.9)$$

where u is a scalar parameter, $-\infty < u < \infty$, and $0 \le a \le T$. Let

$$\varphi(u, a) = \min_y K(y, u, a). \qquad (8.18.10)$$

To obtain an equation for $\varphi(u, a)$, we write

$$\varphi(u, a) = \min_y \left\{ b_1 \left[u + (b(a), y(a))\Delta + \int_{a+\Delta}^T (b(t_1), y(t_1)) \, dt_1 \right]^2 \right.$$

$$\left. + (y(a), y(a))\Delta + \int_{a+\Delta}^T (y, y) \, dt \right\} + O(\Delta^2)$$

$$= \min_{y(a)} [(y(a), y(a))\Delta + \varphi(u + (b(a), y(a))\Delta, a + \Delta)] + O(\Delta^2).$$

$$(8.18.11)$$

Hence, letting $\Delta \to 0$, we obtain

$$-\frac{\partial \varphi}{\partial a} = \min_v \left[(v, v) + (b(a), v) \frac{\partial \varphi}{\partial u} \right]. \qquad (8.18.12)$$

Setting $\varphi(u) = u^2 \rho(a)$, we readily find the Riccati equation for $\rho(a)$. The initial condition is $\rho(T) = b_1$.

EXERCISES

1. Carry through the determination of $\rho(a)$ and show how to use it to obtain the solution of the original problem, the minimization of $J(x, y)$.

2. Carry through the details of the solution when $k > 1$.

3. Use the foregoing method to minimize

$$J(u, v) = a_1 u(T)^2 + a_2 u'(T)^2 + \cdots + a_k u^{(k)}(T)^2 + \int_0^T v^2 \, dt,$$

where

$$u^{(n)} + b_1 u^{(n-1)} + \cdots + b_n u = v, \quad u^{(i)}(0) = c_i, \quad i = 0, 1, \ldots, N - 1.$$

4. What modifications are required if there are some terminal conditions, $u^{(j)}(T) = d_i$, $i = 1, 2, \ldots, r$?

5. Carry out the details of the solution for the discrete control process where $x_{n+1} = Ax_n + y_n$, $x_0 = c$,

$$J(x, y) = (x(N), Bx(N)) + \sum_{k=0}^{N} (y(N), y(N)).$$

6. Carry through the details of the solution for the case where

$$J(x, y) = (x(T), Bx(T)) + (x(T), b) + \int_0^T (y, y) \, dt$$

with B of the form shown in (8.18.4) and $b = (b_1, b_2, \ldots, b_k, 0, \ldots, 0)$.

8.19. Successive Approximations

The purpose of successive approximations is to approach the solution of some particular problem through a sequence of problems that can be resolved in terms of available algorithms, analytic or computational. For this reason, the linear equation plays a dominant role in classical analysis, and will continue to play a major role in all subsequent analysis. The linearity of the equation makes it easy for us to construct manageable algorithms for both numerical solution and qualitative investigations.

With the advent of the digital computer the number of algorithms that can truthfully be called efficient has increased vastly. What is particularly significant is that in many important cases we no longer constrained to use approximations based on linear equations. Consequently, in place of focussing solely on approximations of this restricted nature, we can concentrate on algorithms that are feasible in terms of available computers, analog, digital, hybrid, man-machine, and so on.

In particular, we can fasten our attention on a new type of mathematical problem, that of constructing a sequence of low-dimensional approximations that will converge to the solution of a high-dimensional problem. Dynamic programming is a theory devoted to this objective.

Consider, for example, the problem of minimizing

$$J(x, y) = (x(T), Bx(T)) + \int_0^T (y, y)\, dt, \qquad (8.19.1)$$

where

$$x' = Ax + y, \qquad x(0) = c, \qquad (8.19.2)$$

and B is an $N \times N$ matrix. Let us choose an initial approximation y_0 and determine x_0 from (8.19.2) in this fashion. Let us use the values $x_{k+1}(T), \ldots, x_N(T)$ obtained from $x_0(T)$ in (8.19.1) so that

$$J(x, y) = (x(T), B_1 x(T)) + (b, x(T)) + \int_0^T (y, y)\, dt + h_0, \qquad (8.19.3)$$

where B_1 is a new matrix having the form shown in (8.18.4), b has the form $(b_1, b_2, \ldots, b_k, 0, \ldots, 0)$ and h_0 is a scalar quantity calculated using the values $x_{k+1}(T), \ldots, x_N(T)$.

This new variational problem can now be approached in the fashion described in Section 8.18. This procedure can then be iterated in various fashions to obtain a method of successive approximations. The convergence of methods of this type will be examined in Volume II.

Once we have a systematic method for the minimization of $J(x, y)$, we can consider the problem of minimizing a more general functional of the form

$$K(x, y) = g(x(T)) + \int_0^T h(y)\, dt. \qquad (8.19.4)$$

Using the quadratic approximation to $g(x(T))$ and $h(y)$ we can construct a sequence of approximate variational problems of the type appearing in (8.19.1).

Furthermore, we can attack the problem of minimizing the more frequently appearing cost functional

$$L(x, y) = g(x(T)) + \int_0^T h(x, y)\, dt \qquad (8.19.5)$$

by means of successive approximations based upon

$$L_n(x, y) = g(x(T)) + \int_0^T h(x_n, y)\, dt \qquad (8.19.6)$$

and use of the preceding ideas.

The foregoing discussion, brief as it is, is intended to highlight two basic ideas. The first is that we can sensibly contemplate hierarchies of methods of successive approximations with the computer at our disposal; the second is that this same device allows us a flexibility in the use of successive approximations that is, to say the least, exhilarating. All that is required is boldness and imagination on the part of the mathematician.

8.20. Distributed Parameters

A new and interesting direction of research in control theory centers around processes described by partial differential equations, differential-difference equations, and still more general functional equations. For example, we may encounter the problem of minimizing

$$J(u, v) = \sum_{i=1}^{k} b_i u(x_i, T)^2 + \int_0^T \int_0^1 v(x, t_1)^2 \, dx \, dt_1, \quad (8.20.1)$$

where

$$\begin{aligned} u_t &= u_{xx} + v, \quad & 0 < x < 1, \quad & t > 0, \\ u(0, t) &= u(1, t), \quad & & t > 0, \quad (8.20.2) \\ u(x, 0) &= g(x), \end{aligned}$$

or that of minimizing

$$J(u, v) = \sum_{i=1}^{k} b_i u(T - \tau_i)^2 + \int_1^T v^2 \, dt_1, \quad (8.20.3)$$

where

$$\begin{aligned} u'(t) &= a_1 u(t) + a_2 u(t - 1) + v, \quad & t \geq 1, \\ u(t) &= g(t), \quad & 0 \leq t \leq 1. \end{aligned} \quad (8.20.4)$$

This type of problem may be treated directly by means of functionals of the initial functions, leading to functional-differential equations that present considerable difficulties as far as numerical solution is concerned. The foregoing methods can, however, be applied to provide feasible algorithms.

This is a point of some importance in connection with control processes ruled by partial differential equations since a cost functional such as $\int_0^1 g(u(x, T))\, dx$ can always be well approximated to by a finite sum using a quadrature formula,

$$\int_0^1 g(u(x, T))\, dx \cong \sum_{i=1}^{k} w_i\, g(u(x_i, T)). \qquad (8.20.5)$$

8.21. Slightly Intertwined Systems

Let us now indicate another way in which the analytic structure of the problem can be exploited to provide a formulation in terms of functions of low dimension.

In many physical problems we encounter matrices that have the form

$$B = \begin{pmatrix} B_1 & & & & 0 \\ & B_2 & & & \\ & & \cdot & & \\ & & & \cdot & \\ 0 & & & & \cdot \\ & & & & B_N \end{pmatrix}, \qquad (8.21.1)$$

a block-diagonal matrix. The study of the behavior of the system under examination can then be reduced to the study of N distinct subsystems governed by the matrices $B_1, B_2, \ldots, B_N$. The subsystems are said to be uncoupled.

In many other situations there is weak coupling between the individual subsystems. One example of this is that where B is nearly block-diagonal with only an occasional off-diagonal term. It is reasonable to assume that there should be some systematic procedure for taking advantage of this characteristic property.

To illustrate the kind of method that can be employed, let us consider a specific problem, the determination of the solution of a set of linear equations of the form

$$a_{11}x_1 + a_{12}x_2 + a_{13}x_3 = c_1$$
$$a_{21}x_1 + a_{22}x_2 + a_{23}x_3 = c_2$$
$$a_{31}x_1 + a_{32}x_2 + a_{33}x_3 + b_1x_4 = c_3$$
$$b_1x_3 + a_{44}x_4 + a_{45}x_5 + a_{46}x_6 = c_4$$
$$a_{54}x_4 + a_{55}x_5 + a_{56}x_6 = c_5$$
$$a_{64}x_4 + a_{65}x_5 + a_{66}x_6 + b_2x_7 = c_6$$
$$\vdots$$

$$(8.21.2)$$

$$b_{N-1}x_{3N-3} + a_{3N-2,3N-2}x_{3N-2} + a_{3N-2,3N-1}x_{3N-1}$$
$$+ a_{3N-2,3N}x_{3N} = c_{3N-2}$$
$$a_{3N-1,3N-2}x_{3N-2} + a_{3N-1,3N-1}x_{3N-1}$$
$$+ a_{3N-1,3N}x_{3N} = c_{3N-1}$$
$$a_{3N,3N-2}x_{3N-2} \quad \begin{aligned}&+ a_{3N,3N-1}x_{3N-1}\\&+ a_{3N,3N}x_{3N} = c_{3N}.\end{aligned}$$

If the coefficients b_i were all zero, this problem would decompose into simple three-dimensional problems. To simplify the subsequent analysis, introduce the matrices

$$A_k = (a_{i+3k-3,j+3k-2}), \qquad i, j = 1, 2, 3, \qquad (8.21.3)$$

$k = 1, 2, \ldots,$ and the vectors

$$x^k = \begin{pmatrix} x_{3k-2} \\ x_{3k-1} \\ x_{3k} \end{pmatrix}, \qquad c^k = \begin{pmatrix} c_{3k-2} \\ c_{3k-1} \\ c_{3k} \end{pmatrix}. \qquad (8.21.4)$$

We suppose that the matrix of coefficients in (8.21.2) is positive definite. Hence, the solution of the system in (8.21.2) furnishes the minimum of the inhomogeneous quadratic form

$$(x^1, A_1x^1) + (x^2, A_2x^2) + \cdots + (x^N, A_Nx^N)$$
$$- 2(c^1, x^1) - 2(c^2, x^2) - \cdots - 2(c^N, x^N)$$
$$+ 2b_1x_3x_4 + 2b_2x_6x_7 + \cdots + 2b_{N-1}x_{3N-3}x_{3N-2}. \quad (8.21.5)$$

Let us now use the functional equation technique. Introduce the sequence of functions $\{f_N(z)\}$, $-\infty < z < \infty$, $N = 1, 2, \ldots$, defined by the relation

$$f_N(z) = \min_{x_i} \left[\sum_{i=1}^{N} (x^i, A_i x^i) - 2 \sum_{i=1}^{N} (c^i, x^i) + 2 \sum_{i=1}^{N-1} b_i x_{1+3i} x_{3i} + 2z x_{3N} \right].$$

(8.21.6)

The principle of optimality yields the recurrence relation

$$f_N(z) = \min_{R_N} \left[\min_{x_{3N}, x_{3N-1}} \left[(x^N, A_N x^N) + 2z x_{3N} - 2(c^N, x^N) \right] \right.$$
$$\left. + f_{N-1}(b_{N-1} x_{3N-2}) \right], \qquad (8.21.7)$$

where R_N is the three-dimensional region

$$-\infty < x_{3N}, x_{3N-1}, x_{3N-2} < \infty.$$

If we introduce the functions

$$g_N(z, y) = \min_{x_{3N}, x_{3N-1}} \left[(x^N, A^N x^N) + 2z x_{3N} - 2(c^N, x^N) \right], \qquad (8.21.8)$$

where $x_{3N-2} = y$, the relation in (8.21.7) becomes

$$f_N(z) = \min_{y} \left[g_N(x, y) + f_{N-1}(b_{N-1} y) \right], \qquad (8.21.9)$$

$N = 1, 2, \ldots$.

We can now go further and set

$$f_N(z) = u_N + 2v_N z + w_N z^2, \qquad (8.21.10)$$

and obtain recurrence relations for the sequence $\{u_N, v_N, w_N\}$.

EXERCISES

1. Obtain the recurrence relations and show how they can be used to solve the linear system in (8.21.2).
2. Consider the case where the coupling is k-dimensional rather than one-dimensional.
3. What can be done if the matrix of coefficients is no longer positive definite?

Miscellaneous Exercises

1. Consider the problem of determining the maximum and minimum of the quadratic forms

   (a) $(ax_1)^2 + (x_1 + ax_2)^2 + \cdots + (x_1 + x_2 + \cdots + x_{N-1} + ax_N)^2$,

   (b) $x_1^2 + (x_1 + ax_2)^2 + \cdots + (x_1 + ax_2 + a^2x_3 + \cdots + a^{N-1}x_N)^2$,

   (c) $x_1^2 + (x_1 - ax_2)^2 + \cdots + (x_1 + ax_2 + (a + b)x_3 + \cdots$
   $$+ [a + (N - 2)b]x_N)^2,$$

   all subject to $x_1^2 + x_2^2 + \cdots + x_N^2 = 1$. (For a discussion of these questions by classical methods, see

   A. M. Ostrowski, "On the Bounds for a One-Parameter Family of Matrices," *J. Math.*, **200**, 1958, pp. 190–200.)

2. Show that the Newton-Raphson-Kantorovich approximation $u_n' = 2u_n u_{n-1} - u_{n-1}^2 + a(t)$, $u_n(0) = c$, $n = 1, 2, \ldots$, can be regarded as a particular type of approximation in policy space, starting from $u' = \max_v (2uv - v^2 + a(t))$.

3. Show

$$\frac{d}{dt}\left(P(t)\frac{dx}{dt}\right) + Q(t)x = y,$$

   $a < t < 1$, $x(a) = c$, $x'(1) + Bx(1) = 0$, is the Euler equation associated with

$$J(x, y) = \int_a^1 [(Qx, x) - (Px', x') - 2(y, x)]\, dt - (P(1)Bx(1), x(1)),$$

   where x is subject to the condition $x(a) = c$. Here we assume that P, Q, and y are continuous in $[a, 1]$ and $P(t)$ is nonsingular in $[a, 1]$; P, Q, and $P(1)B$ are taken to be symmetric matrices.

4. Show that

$$-f_a = \tfrac{1}{4}(P^{-1}(a)f_c, f_c) - 2(c, y(a)) + (Q(a)c, c),$$

   where $f(a, c) = \max_x J(x, y)$.

5. Let $\Phi(t)$ be the matrix solution of

$$\frac{d}{dt}\left(P(t)\frac{d\varphi}{dt}\right) + Q\varphi = 0, \qquad \varphi(a) = 1, \qquad \varphi'(1) + B\varphi(1) = 0,$$

and x_0 satisfy the equation in Exercise 3 with $c = 0$. Then

$$f(a, c) = -\int_a^1 (y, x_0)\, dt - \int_a^1 (y, \varphi c)\, dt + (P(a), x_0'(a), c)$$
$$+ (P(a)\varphi'(a)c, c).$$

6. Hence, if $K(t, s, a)$ is the Green's function defined by x_0,

$$x_0(t) = \int_a^1 K(t, s, a)y(s)\, ds,$$

show that $\partial K/\partial a\,(t, s, a) = \varphi(t)P^{-1}(a)\varphi^*(s)$. (Here φ^* denotes the transpose of φ.)

7. Hence show that

$$\frac{\partial K}{\partial a}(t, s, a) = \left[\frac{\partial K}{\partial s}(t, s, a)\right]_{s=a} P(a)\left[\frac{\partial K}{\partial t}(t, s, a)\right]_{t=a}.$$

(For the results of the last five exercises, see

R. Bellman and S. Lehman, "Functional Equations in the Theory of Dynamic Programming—X: Resolvents, Characteristic Functions, and Values," *Duke Math. J.*, **27**, 1960, pp. 55–69.)

8. Let $A = (a_{ij})$ be a Jacobi matrix, by which we mean that $a_{ij} = 0$, $|i - j| \geq 2$,

$$A = \begin{pmatrix} a_{11} & a_{12} & 0 & \cdots \\ a_{21} & a_{22} & a_{23} & 0 \\ 0 & & & \\ \vdots & & & \end{pmatrix},$$

and suppose that it is positive definite. Consider the problem of solving $Ax = c$ by minimizing $Q(x) = (x, Ax) - 2(c, x)$. Here

$$Q(x) = a_{11}x_1^2 + \cdots + a_{NN}x_N^2 - 2c_1x_1 - 2c_2x_2 - \cdots - 2c_N x_N.$$

Consider the more general problem of minimizing

$$Q_k(x, z) = a_{11}x_1{}^2 + \cdots + a_{kk}x_N{}^2 - 2c_1x_1 - 2c_2x_2 - \cdots - 2zx_k.$$

Define $f_k(z) = \min_x Q_k(x, z)$. Then $f_1(z) = z^2/a_{11}$,

$$f_k(z) = \min_{x_k} [a_{kk}x_k{}^2 - 2zx_k + f_{k-1}(c_{k-1} + a_{k-1,k}x_k)],$$

for $k = 2, 3, \ldots, N$. Show that $f_k(z) = u_k + v_k z + w_k z^2$, where u_k, v_k, w_k are independent of z and derive recurrence relations connecting u_k, v_k, w_k with $u_{k-1}, v_{k-1}, w_{k-1}$.

9. Under what conditions on the a_{ij} is the solution $A^{-1}c$ obtained in this fashion valid for general nonsingular A? (*Hint:* The recurrence relations involve analytic functions of the a_{ij}.)
(For a discussion of analytic continuation, see p. 147 et seq. of

R. Bellman, *Introduction to Matrix Analysis*, McGraw-Hill, New York, 1960.

For the connection with standard Gauss method of elimination for the solution of linear algebraic equations, see

S. Lehman, "Dynamic Programming and Gaussian Elimination," *J. Math. Anal. Appl.*, 5, 1962, pp. 499–501.)

10. Consider the equation $Ax = y$, where A is a general positive definite matrix. Let $Q(x) = (x, Ax) - 2(x, y)$. Write

$$f_k(y_1, y_2, \ldots, y_k) = \min_{x_i} \left[\sum_{i,j=1}^{k} a_{ij}x_i x_j - 2\sum_{i=1}^{k} x_i y_j \right]$$

for $k = 1, 2, \ldots, N$. Show that

$$f_k(y_1, y_2, \ldots, y_k)$$
$$= \min_{x_k} [a_{kk}x_k{}^2 - 2x_k y_k + f_{k-1}(y_1 - a_{1k}y_k, \ldots, y_{k-1} - a_{k-1,k}x_k)],$$

and thus obtain a recurrence relation for the matrix R_k defined by $f_k(y_1, y_2, \ldots, y_k) = (y, R_k y)$.

11. To treat the minimization of

$$Q_{N,M}(x) = \sum_{k=0}^{N} \left(b_k - \sum_{l=0}^{M} x_l a_{k-l} \right)^2, \qquad N > M \geq 1,$$

over y, where $x' = Ax + y$, $x(0) = c$. Suppose in place of the state variable c, we introduce the observables $x_1(0)$, $x_1'(0)$, ..., $x_1^{(N-1)}(0)$. Obtain the analytic representation of $f(c, T) = \min_y J(x, y)$ in terms of these new "convenient state variables." Generally, suppose we attempt to introduce the variables $x_1(0)$, $x_1'(0)$, ..., $x_1^{(k)}(0)$, $x_2(0)$, $x_2'(0)$, ..., $x_2^{(l)}(0)$, ... Under what conditions can this be done?

15. Use the functional equation approach to treat the minimization of the Selberg quadratic form

$$\sum_{k \leq N} \left(\sum_{k|N} x_k \right)^2$$

over all x_k subject to $x_1 = 1$.
(See Appendix B of

R. Bellman, *Introduction to Matrix Analysis*, McGraw-Hill, New York, 1960,

and

R. Bellman, "Dynamic Programming and the Quadratic Form of Selberg," *J. Math. Anal. Appl.*, **15**, 1966, pp. 30–32.)

16. Consider the terminal control problem where it is desired to minimize

$$J(x, y) = (x(T), Bx(T)) + \int_0^T [(x, x) + (y, y)]\, dt,$$

where $x' = Ax + y$, $x(0) = c$.

17. Consider the discrete version where

$$J(x, y) = (x_N, Bx_N) + \sum_{k=0}^N [(x_k, x_k) + (y_k, y_k)]$$

and $x_{k+1} = Ax_k + y_k$, $x_0 = c$.

18. Prove by a direct calculation that the solution to the minimization of

$$J(x) = \int_0^T [(x', x') + (x, Ax)]\, dt, \qquad x(0) = c,$$

may be obtained by taking the limit as $\Delta \to 0$ of the solution of the problem of minimizing

$$\sum_{k=0}^N [(x_{k+1} - x_k, x_{k+1} - x_k) + (x_k, Ax_k)\Delta], \qquad x_0 = c.$$

BIBLIOGRAPHY AND COMMENTS

8.1. The exploitation of the fact that the minimization of a quadratic functional leads to a Riccati equation in the continuous case and continued fractions in the discrete case has been systematically carried out by a number of authors. See, for example, for some of the early work,

D. Adorno, *The Asymptotic Theory of Control Systems—I: Stochastic and Deterministic Processes*, Tech. Rept. 32-21, California Inst. of Technol., 1960.

R. Beckwith, *Analytic and Computational Aspects of Dynamic Programming Processes of High Dimension*, Ph.D. Thesis, Mathematics, Purdue Univ., Lafayette, Indiana, 1959.

M. Freimer, "A Dynamic Programming Approach to Adaptive Control Processes," *IRE Trans. Auto. Control*, **AC-4**, 1959, pp. 10–15.

J. D. R. Kramer, "On Control of Linear Systems with Time Lags," *Information and Control*, **3**, 1960, pp. 299–326.

(The foregoing were Ph.D. theses, published and unpublished versions, written at various universities under the author's supervision.)

R. E. Kalman and R. W. Koepcke, "The Role of Digital Computers in the Dynamic Optimization of Chemical Reactions," *Proc. Western Joint Computer Conference*, San Francisco, 1959, pp. 107–116.

R. E. Kalman, L. Lapidus, and E. Shapiro, "The Optimal Control of Chemical and Petroleum Processes," *Proc. Joint Symposium on Instrumentation and Computation in Process Development and Plant Design*, Inst. Chem. Eng., London, 1961, pp. 6–17.

J. T. Tou, "Design of Optimum Digital Control Systems via Dynamic Programming," *Proc. Dynamic Programming Workshop*, J. E. Gibson, editor, Purdue Univ., Lafayette, Indiana, 1961, pp. 37–66.

J. T. Tou, *Optimum Design of Digital Control Systems*, Academic Press, New York, 1963.

The first observation of this was made in

R. Bellman, "Some New Techniques in the Dynamic Programming Solution of Variational Problems," *Quart. Appl. Math.*, **16**, 1958, pp. 295–305.

Some books with more detailed discussions are

R. Bellman, *Introduction to Matrix Analysis*, McGraw-Hill, New York, 1960.

The concept of imbedding a particular problem in a family of related problems can be carried over to the study of processes where no control is involved. This constitutes the theory of invariant imbedding.

For some detailed discussion and applications, see

R. Bellman, *Adaptive Control Processes: A Guided Tour*, Princeton Univ. Press, Princeton, New Jersey, 1961.

R. Bellman and R. Kalaba, *Dynamic Programming and Modern Control Theory*, Academic Press, New York, 1966.

R. Bellman, R. Kalaba, and M. Prestrud, *Invariant Imbedding and Radiative Transfer in Slabs of Finite Thickness*, American Elsevier, New York, 1963.

R. Bellman, H. Kagiwada, R. Kalaba, and M. Prestrud, *Invariant Imbedding and Time-Dependent Processes*, American Elsevier, New York, 1964.

R. Bellman, R. Kalaba, and G. M. Wing, "Invariant Imbedding and Mathematical Physics—I: Particle Processes," *J. Math. Phys.*, **1**, 1960, pp. 280–308.

R. Bellman and R. Kalaba, "On the Fundamental Equations of Invariant Imbedding —I," *Proc. Nat. Acad. Sci. U.S.A.*, **47**, 1961, pp. 336–338.

R. Bellman, R. Kalaba, and G. M. Wing, "Invariant Imbedding and the Reduction of Two-Point Boundary Value Problems to Initial Value Problems," *Proc. Nat. Acad. Sci. U.S.A.*, **46**, 1960, pp. 1646–1649.

R. Bellman and R. Kalaba, "A Note on Hamilton's Equations and Invariant Imbedding," *Quart. Appl. Math.*, **21**, 1963, pp. 166–168.

R. Bellman and R. Kalaba, "Invariant Imbedding and the Integration of Hamilton's Equations," *Rend. Circ. Mat. Palermo*, **12**, 1963, pp. 1–11.

R. Bellman and T. Brown, "A Note on Invariant Imbedding and Generalized Semi-Groups," *J. Math. Anal. Appl.*, **9**, 1964, pp. 394–396.

R. Bellman, H. Kagiwada, R. Kalaba, and R. Sridhar, "Invariant Imbedding and Nonlinear Filtering Theory," *J. Astro. Sci.*, **13**, 1966, pp. 110–115.

R. Bellman, H. Kagiwada, and R. Kalaba, "Invariant Imbedding and Nonvariational Principles in Analytical Dynamics," *Internat. J. Nonlinear Mech.*, **1**, 1965, pp. 51–55.

8.3. See the bibliography of papers by Reid and others given at the end of Chapter 7.

8.7. Special-purpose digital computers specifically designed to solve systems of ordinary differential equations with initial conditions may soon render discussions of this type unnecessary. In view of the frequent occurrence of Riccati-type differential equations in mathematical physics (see

R. Bellman, R. Kalaba, and M. Prestrud, *Invariant Imbedding and Radiative Transfer in Slabs of Finite Thickness*, American Elsevier, New York, 1963.

R. Bellman, H. Kagiwada, R. Kalaba, and M. Prestrud, *Invariant Imbedding and Time-Dependent Processes*, American Elsevier, New York, 1964),

it may be worthwhile to build special-purpose computers solely for the purpose of solving equations of this type. In any case, the point we wish to emphasize is that the effective numerical solution of high-dimensional equations requires novel concepts and techniques.

8.11. In Section 8.21, other types of partitioning are discussed. In general, this is a new problem area in analysis that has not been systematically investigated.

8.13. For an application of extrapolation methods, where numerous other references may be found, see

R. Bellman and R. Kalaba, "A Note on Nonlinear Summability Techniques in Invariant Imbedding," *J. Math. Anal. Appl.*, **6**, 1963, pp. 465–472.

R. Bellman, H. Kagiwada, and R. Kalaba, "Nonlinear Extrapolation and Two-Point Boundary-Value Problems," *Comm. ACM*, **8**, 1965, pp. 511–512.

8.14. It cannot be overemphasized that the same variational problem may profitably be attacked in a number of different ways. In this volume we have omitted any discussion of Rayleigh-Ritz techniques or of the gradient method.

For an extensive treatment of inequalities involving quadratic forms and functionals, see

E. F. Beckenbach and R. Bellman, *Inequalities*, Springer, Berlin, 1961.

8.16. For further results, see

R. Bellman, R. Kalaba, and J. Lockett, *Numerical Inversion of the Laplace Transform with Applications*, American Elsevier, New York, 1966.

8.17. See

R. Bellman, "Dynamic Programming and Lagrange Multipliers," *Proc. Nat. Acad. Sci. U.S.A.*, **42**, 1956, pp. 767–769.

8.18. See

R. Bellman, "Terminal Control, Time Lags, and Dynamic Programming," *Proc. Nat. Acad. Sci. U.S.A.*, **43**, 1957, pp. 927–930.
R. Bellman and R. Kalaba, "Reduction of Dimensionality, Dynamic Programming, and Control Processes," *J. Basic Eng.*, March, 1961, pp. 82–84.
R. Bellman, "On the Reduction of Dimensionality for Classes of Dynamic Programming Processes," *J. Math. Anal. Appl.*, **3**, 1961, pp. 358–360.
R. Bellman, "On the Application of Dynamic Programming to the Determination of Optimal Play in Chess and Checkers," *Proc. Nat. Acad. Sci. U.S.A.*, **53**, 1965, pp. 244–247.
R. Bellman, "Dynamic Programming, Generalized States, and Switching Systems," *J. Math. Anal. Appl.*, **12**, 1965, pp. 360–363.

8.20. See

R. Bellman and R. Kalaba, "Dynamic Programming Applied to Control Processes Governed by General Functional Equations," *Proc. Nat. Acad. Sci. U.S.A.*, **48**, 1962, pp. 1735–1737.
A. G. Butkovskii, *Optimal Control Theory for Distributed Parameter Systems*, Science Publishing House, Moscow, 1965; English translation, American Elsevier, New York, 1967. To appear.

8.21. See

R. Bellman, "On Some Applications of Dynamic Programming to Matrix Theory," *Illinois J. Math.*, **1**, 1957, pp. 297–301.
R. Bellman, "On the Computational Solution of Linear Programming Problems Involving Almost Block Diagonal Matrices," *Management Sci.*, **3**, 1957, pp. 403–406.
G. B. Dantzig, "On the Status of Multistage Linear Programming Problems," *Management Sci.*, **6**, 1959, pp. 53–72.
G. B. Dantzig and P. Wolfe, "Decomposition Principles for Linear Programs," *Operations Res.*, **8**, 1960, pp. 101–111.

9

FUNCTIONAL ANALYSIS

9.1. Motivation

Up until this point, we have concentrated upon the most familiar functional equations of classical analysis, ordinary differential equations. Furthermore, our evaluation of the utility of a control process was in terms of well-behaved quadratic functionals. The difficult task of an a priori establishment of the existence of an extremal function, or equivalently an optimal policy, was avoided by following a carefully chosen circuitous route.

First the Euler equation was obtained using techniques of dubious validity. Then the linearity of the equation was exploited to establish the existence of a solution of this equation and then to demonstrate its uniqueness. Finally, the quadratic nature of the functional was used to show that the function so obtained provided an absolute minimum.

With the proper analytic background, we can use the same approach to study more general control processes where the system is governed by more complex functional equations such as

$$x'(t) = Ax(t) + Bx(t - 1) + y(t) \qquad (9.1.1)$$

or

$$u_t = u_{xx} + v(x, t), \qquad (9.1.2)$$

provided that the criterion function is still a quadratic functional.

There are, however, several motives for pursuing a different approach. In the first place, the deviousness described above is esthetically displeasing. In the second place, we wish to lay the frame-work for the study of more general control processes involving nonlinear equations and constraints.

A basic tool for these purposes is the theory of functional analysis. For our present purposes, it is sufficient to remain within the well-charted domain of the Hilbert space $L^2(0, T)$, where T is finite.

We will begin by recalling some of the fundamental concepts of Hilbert space and then apply these ideas to the basic variational problem we have been considering throughout the previous chapters. Following this, we will turn our attention to the analytic aspects of solving the Euler equation and finally touch on some of the computational problems.

9.2. The Hilbert Space $L^2(0, T)$

Consider the set of real scalar functions $f(t)$ defined over $[0, T]$ and possessing the property that $\int_0^T f^2 \, dt$ exists as a Lebesgue integral. Call this space $L^2(0, T)$ and write $f \in L^2(0, T)$. This is a linear space in the sense that $f + g$ belongs to the space whenever f and g do, but it is not finite-dimensional.

We introduce a metric by means of the distance formula

$$d(f, g) = \left(\int_0^T (f - g)^2 \, dt \right)^{1/2}. \tag{9.2.1}$$

This possesses the usual attributes of a distance, namely, that $d(f, g) = d(g, f)$, that $d(f, g) > 0$ for $f \neq g$, that $d(f, g) = 0$ implies that $f = g$,† and finally that the triangle inequality holds,

$$d(f, g) \leq d(f, h) + d(h, g), \tag{9.2.2}$$

for any three functions f, g, and h in $L^2[0, T]$.

It is customary to use the norm symbol

$$\|f - g\| = \left(\int_0^T (f - g)^2 \, dt \right)^{1/2} \tag{9.2.3}$$

in place of the distance function.

---

† All equations in $L^2(0, T)$ hold almost everywhere, which is to say apart from a set of measure zero.

Once having introduced a metric, we can talk about the convergence of a sequence of functions $\{f_n\}$. If there exists a function $f \in L^2(0, T)$ such that

$$\|f_n - f\| \to 0 \tag{9.2.4}$$

as $n \to \infty$, we say that f is the limit of the sequence $\{f_n\}$. This is called "strong convergence" to distinguish it from another type of convergence that is occasionally useful.

One of the basic results in Hilbert space is that a necessary and sufficient condition for strong convergence is that $\|f_n - f_m\| \leq \varepsilon$ for $m, n \geq n_0(\varepsilon)$.

9.3. Inner Products

If $f, g \in L^2(0, T)$, the inequality

$$|fg| \leq \frac{f^2 + g^2}{2} \tag{9.3.1}$$

shows that the integral $\int_0^T fg\, dt$ exists. This is called the inner product and we introduce the notation

$$[f, g] = \int_0^T fg\, dt. \tag{9.3.2}$$

It is clear that

(a) $[f, g] = [g, f]$,

(b) $[f, g + h] = [f, g] + [f, h]$. $\tag{9.3.3}$

The Cauchy-Schwarz inequality may be written

$$[f, g]^2 \leq [f, f][g, g]. \tag{9.3.4}$$

EXERCISES

1. We say that $\{f_n\}$ converges weakly to a function $f \in L^2(0, T)$ if $\int_0^T f_n g\, dt \to \int_0^T fg\, dt$ for every $g \in L^2(0, T)$. Show that strong convergence implies weak convergence, but not conversely.

2. Show that the existence of the integral $\int_0^T fg\, dt$ for all $g \in L^2(0, T)$ implies that $f \in L^2(0, T)$.

9.4. Linear Operators

An operator is a transformation from a function space to a function space. We are primarily interested in linear operators of the form

$$Af = \int_0^T k(t, t_1) f(t_1) \, dt_1, \tag{9.4.1}$$

where k is such that $Af \in L^2(0, T)$ whenever $f \in L^2(0, T)$. Actually, in our present work, we enjoy the stronger property that $k(t, t_1)$ is continuous over $0 \le t, t_1 \le T$. The adjoint operator A^* is defined by the relation

$$[Af, g] = [f, A^*g]. \tag{9.4.2}$$

In the case where (9.4.1) holds, we have

$$\int_0^T (Af)g \, dt = \int_0^T \left(\int_0^T k(t, t_1) f(t_1) \, dt_1 \right) g(t) \, dt$$

$$= \int_0^T \left(\int_0^T k(t, t_1) g(t) \, dt \right) f(t_1) \, dt_1. \tag{9.4.3}$$

Hence,

$$A^*g = \int_0^T k(t, t_1) g(t) \, dt. \tag{9.4.4}$$

The interchange of orders of integration is legitimate if k is bounded. The operator is called self-adjoint if $A = A^*$.

The self-adjoint operator A is called " positive definite " if

$$[Af, f] = 0 \tag{9.4.5}$$

implies that $f = 0$.

9.5. Vector Hilbert Space

We will be dealing with the vector Hilbert space $L^2(0, T)$ defined by the condition that $\int_0^T (f, f) \, dt$ exists. Here (f, f) is the vector inner product used in the previous chapter. We shall employ the same notation,

$$[f, f] = \int_0^T (f, f) \, dt. \tag{9.5.1}$$

The linear operators of interest to us have the form

$$Af = \int_0^T K(t, t_1) f(t_1)\, dt_1, \tag{9.5.2}$$

where K is a bounded matrix for $0 \le t, t_1 \le T$. The adjoint operator is

$$A^* g = \int_0^T K(t, t_1)' g(t)\, dt, \tag{9.5.3}$$

where K' is the transpose matrix.

9.6. Quadratic Functionals

Let A be as in (9.5.2), a linear operator with the property that $Af \in L^2(0, T)$ when $f \in L^2(0, T)$. We introduce the quadratic functional

$$J_a(f) = [g + Af, g + Af] + a[f, f], \tag{9.6.1}$$

where a is a positive constant, and $f, g \in L^2(0, T)$.

The motivation for this is the following. Consider the control process studied in Chapter 8. We wish to minimize the functional

$$J(x, y) = \int_0^T [(x, x) + (y, y)]\, dt, \tag{9.6.2}$$

where x and y are connected by the linear differential equation

$$\frac{dx}{dt} = Bx + y, \qquad x(0) = c. \tag{9.6.3}$$

We can eliminate x by solving (9.6.3),

$$x = e^{Bt} c + \int_0^t e^{B(t - t_1)} y(t_1)\, dt_1. \tag{9.6.4}$$

Hence, the problem is that of minimizing

$$K(y) = \int_0^T \left[\left(e^{Bt} c + \int_0^t e^{B(t - t_1)} y(t_1)\, dt_1,\ e^{Bt} c + \int_0^t e^{B(t - t_1)} y(t_1)\, dt_1 \right) \right.$$
$$\left. + (y, y) \right] dt. \tag{9.6.5}$$

If we write

$$g = e^{Bt}c, \qquad Ay = \int_0^t e^{B(t-t_1)}y(t_1)\,dt_1, \qquad (9.6.6)$$

the problem is that of minimizing

$$K(y) = [g + Ay, g + Ay] + [y, y]. \qquad (9.6.7)$$

EXERCISE

1. What is the operator adjoint to A as defined in (9.6.6)?

9.7. Existence and Uniqueness of a Minimizing Function

Our aim is to establish the following basic result.

Theorem. *There is a unique $f \in L^2(0, T)$ that furnishes the minimum value of $J_a(f)$.*

The proof is divided into two parts. In the first part, the existence of a minimizing function is established and in the second part the uniqueness is demonstrated.

Since $f = 0$ is an admissible function and $J_a(0) = [g, g]$, it is clear that we need only consider functions f for which $J_a(f) \le [g, g]$. Let

$$\alpha = \inf_f J_a(f), \qquad (9.7.1)$$

and $\{f_n\}$ be a sequence with the property that $\lim n \to \infty\ J_a(f_n) = \alpha$. We wish to show that f_n converges strongly to an element f in L^2.

Write

$$K(f) = [Af, Af] + 2[Af, g] + a[f, f], \qquad (9.7.2)$$

so that $J_a(f) = K(f) + [g, g]$.

A straightforward calculation yields

$$K\left(\frac{f_n - f_m}{2}\right) + K\left(\frac{f_n + f_m}{2}\right) = \tfrac{1}{2}[[Af_n, Af_n] + a[f_n, f_n]$$
$$+ [Af_m, Af_m] + a[f_m, f_m]]$$
$$+ 2[Af_n, g]. \qquad (9.7.3)$$

The right-hand side may be written

$$\tfrac{1}{2}[K(f_n) - 2[Af_n, g] + K(f_m) - 2[Af_m, g]] + 2[Af_n, g]. \quad (9.7.4)$$

Hence,

$$K\left(\frac{f_n - f_m}{2}\right) = \frac{1}{2}\left[K(f_n) + K(f_m) - 2K\left(\frac{f_n + f_m}{2}\right)\right]$$

$$+ [Af_n, g] - [Af_m, g]. \quad (9.7.5)$$

Using the expression for $K(f)$ given in (9.7.2), we obtain the desired identity,

$$\left[A\left(\frac{f_n - f_m}{2}\right), A\left(\frac{f_n - f_m}{2}\right)\right] + a\left[\frac{f_n - f_m}{2}, \frac{f_n - f_m}{2}\right]$$

$$= \frac{1}{2}\left[K(f_n) + K(f_m) - 2K\left(\frac{f_n + f_m}{2}\right)\right]. \quad (9.7.6)$$

Since $J_a(f_n) \to \alpha$ as $n \to \infty$, we know that $K(f_n) = J_a(f_n) - [g, g]$ converges to the limit $\alpha - [g, g]$, which we designate by β. Since β is the infimum of $K(f)$ for $f \in L^2[0, T]$, it follows that $K[(f_n + f_m)/2] \geq \beta$. Using this, (9.7.6) yields

$$\left[A\left(\frac{f_n - f_m}{2}\right), A\left(\frac{f_n - f_m}{2}\right)\right] + a\left[\frac{f_n - f_m}{2}, \frac{f_n - f_m}{2}\right]$$

$$\leq \tfrac{1}{2}[K(f_n) + K(f_m) - 2\beta]. \quad (9.7.7)$$

Since $K(f_n)$ and $K(f_m)$ approach β as $m, n \to \infty$, we see that the right-hand side of (9.7.7) approaches zero as $m, n \to \infty$. Since both terms on the left-hand side are nonnegative, each must also approach zero as $m, n \to \infty$. Thus we have the desired results: as $m, n \to \infty$,

$$[f_n - f_m, f_n - f_m] \to 0,$$

$$[A(f_n - f_m), A(f_n - f_m)] \to 0. \quad (9.7.8)$$

It follows that $\{f_n\}$ converges strongly to a function $f_a \in L^2(0, T)$, and likewise $\{Af_n\}$ converges strongly to a function h that must equal Af_a. Hence, $J_a(f_a) = \alpha$; the infimum is actually assumed.

To demonstrate the uniqueness of the minimizing function, we use the convexity of $J_a(f)$. Suppose that f_1 and f_2 were two distinct functions such that

$$J_a(f_1) = \alpha, \qquad J_a(f_2) = \alpha. \qquad (9.7.9)$$

Consider

$$u(\lambda) = J_a(\lambda f_1 + (1 - \lambda)f_2) \qquad (9.7.10)$$

where $0 \leq \lambda \leq 1$. It is clear that $u(\lambda)$ is a quadratic in λ and that

$$u''(\lambda) = 2[A(f_1 - f_2), A(f_1 - f_2)] + a[f_1 - f_2, f_1 - f_2] > 0. \qquad (9.7.11)$$

Hence, $u(\lambda)$ has the form shown in Fig. 9.1. This contradicts the fact that α, the value of $u(\lambda)$ at $\lambda = 0$ and $\lambda = 1$, is the absolute minimum. It follows that f_1 and f_2 cannot be distinct functions.

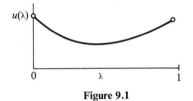

Figure 9.1

EXERCISES

1. Let $L(f)$ be a linear functional on $L^2(0, T)$. By this, we mean that $L(f_1 + f_2) = L(f_1) + L(f_2)$ for any $f_1, f_2 \in L^2(0, T)$ and $|L(f)| \leq k_1 \|f\|$ for $f \in L^2(0, T)$, where k_1 is independent of f. Show that the quadratic functional $Q(f) = \int_0^T f^2 \, dt - 2L(f)$ possesses a unique minimizing function g.
2. Consider the function $u(\varepsilon) = Q(g + \varepsilon h)$ as a function of ε. Using the fact that $u(\varepsilon)$ has an absolute minimum at $\varepsilon = 0$, show that $\int_0^T gh \, dt = L(h)$ for all h. (The Riesz representation theorem.)

9.8. The Equation for the Minimizing Function

Let us now obtain an equation for the minimizing function f_a. Consider

$$J(f_a + \lambda f) = [g + Af_a + \lambda Af, g + Af_a + \lambda Af] + a[f_a + \lambda f, f_a + \lambda f]$$
$$= J_a(f_a) + 2\lambda\{[g + Af_a, Af] + a[f_a, f]\} + \lambda^2\{\cdots\}. \qquad (9.8.1)$$

Since $J(f_a + \lambda f)$ attains its minimum value at $\lambda = 0$, we must have

$$[g + Af_a, Af] + a[f_a, f] = 0 \qquad (9.8.2)$$

for all f. This is the basic variational equation corresponding to the Euler equation derived in the previous chapters.

Using the adjoint operator, (9.8.2) may be written

$$[A^*g + A^*Af_a, f] + a[f_a, f] = 0. \qquad (9.8.3)$$

Since this equation holds for all $f \in L^2(0, T)$, we expect that the condition satisfied by f_a is

$$A^*Af_a + af_a + A^*g = 0. \qquad (9.8.4)$$

This is a valid conclusion, since our assumptions concerning A permit us to conclude that $(A^*Af_a + af_a + A^*g) \in L^2(0, T)$. With this choice of f in (9.8.3), we have

$$[A^*Af_a + af_a + A^*g, A^*Af_a + af_a + A^*g] = 0, \qquad (9.8.5)$$

whence (9.8.4) follows.

We know that (9.8.4) possesses one solution, namely f_a. The question is whether or not this is the unique solution. Suppose that there were two solutions, f_a and f. Then

$$A^*A(f_a - f) + a(f_a - f) = 0, \qquad (9.8.6)$$

and thus

$$[A^*A(f_a - f) + a(f_a - f), f_a - f] = 0. \qquad (9.8.7)$$

Writing this in the form

$$[A^*A(f_a - f), f_a - f] + a[(f_a - f), (f_a - f)] = 0 \qquad (9.8.8)$$

or

$$[A(f_a - f), A(f_a - f)] + a[(f_a - f), (f_a - f)] = 0, \qquad (9.8.9)$$

we obtain a contradiction since both terms on the left are nonnegative and $a[(f_a - f), (f_a - f)] > 0$.

Let us agree to write the unique solution of (9.8.4) in the form

$$f_a = (-a - A^*A)^{-1}A^*g, \qquad (9.8.10)$$

and introduce the notation

$$R_a = (-a - A^*A)^{-1}. \qquad (9.8.11)$$

9.9. Application to Differential Equations

Let us now specialize these results so as to study a control process governed by the differential equation

$$\frac{dx}{dt} = Bx + y, \qquad x(0) = c. \tag{9.9.1}$$

We use B to denote a matrix here in order to avoid confusion with the operator A appearing above. We have

$$x = e^{Bt}c + \int_0^t e^{B(t-s)}y(s)\,ds = z(t) + \int_0^t X(t-s)y(s)\,ds. \tag{9.9.2}$$

Consider the problem of minimizing the functional

$$J(x, y) = \int_0^T [(x - c, x - c) + a(y, y)]\,dt. \tag{9.9.3}$$

This is equivalent to minimizing

$$K(y) = \int_0^T \left[\left(z(t) - c + \int_0^t X(t-s)y(s)\,ds, z(t) - c \right.\right.$$

$$\left.\left. + \int_0^t X(t-s)y(s)\,ds \right) ds + a(y, y) \right] dt. \tag{9.9.4}$$

If we set

$$g = z - c, \qquad Ay = \int_0^t X(t-s)y(s)\,ds, \tag{9.9.5}$$

we see that

$$K(y) = [g - Ay, g - Ay] + a[y, y], \tag{9.9.6}$$

precisely the type of quadratic functional we have been treating.

The adjoint operator is readily seen to be

$$A^*y = \int_t^T X(t-s)'y(s)\,ds, \tag{9.9.7}$$

where X' is the transpose of X.

The function y that minimizes satisfies the equation

$$A^*Ay + ay = -A^*g. \tag{9.9.8}$$

From the expression in (9.9.7) and form of X we see that

$$\left(\frac{d}{dt} - B'\right)A^*w = w \tag{9.9.9}$$

for any w. Hence, (9.9.8) yields

$$a\left(\frac{dy}{dt} - B'y\right) + Ay = -g. \tag{9.9.10}$$

Using the fact that

$$\left(\frac{d}{dt} - B\right)Aw = w \tag{9.9.11}$$

for all w, we see that (9.9.10) yields

$$a\left(\frac{d}{dt} - B\right)\left(\frac{d}{dt} - B'\right)y + y = -\left(\frac{d}{dt} - B\right)g = -Bc. \tag{9.9.12}$$

The boundary conditions are derived from (9.9.8) and (9.9.10), namely,

$$y(T) = 0, \qquad \frac{dy}{dt} - B'y|_{t=0} = 0. \tag{9.9.13}$$

We have thus established directly that there is a unique solution of (9.9.12) satisfying the two-point boundary conditions of (9.9.13).

9.10. Numerical Aspects

The integral equation of (9.9.8) has the form

$$y(t) + \lambda \int_0^T K(t, t_1)y(t_1)\, dt_1 = h(t), \tag{9.10.1}$$

where K is a symmetric kernel. As we know, we can reduce it to a linear differential equation satisfying a two-point boundary condition. In place of doing this, let us examine the use of the Liouville-Neumann series obtained by straightforward iteration,

$$y(t) = h(t) - \lambda \int_0^T K(t, t_1)h(t_1)\,dt_1 + \cdots. \qquad (9.10.2)$$

The radius of convergence of this series is determined by the location of the first characteristic value of the kernel $K(t, t_1)$ and is necessarily finite. From what has preceded, we know that (9.10.1) has a unique solution for $\lambda \geq 0$. It is rather amazing then to find a straightforward determination of the solution of the type given in (9.10.2) thwarted by the fact that the criterion for the original control process is not meaningful if λ is a large negative quantity.

A question of some theoretical and computational importance is whether or not it is possible to obtain a series expansion for the solution of (9.10.1) that will be valid for all $\lambda \geq 0$. As we shall see below, we can accomplish this by expanding the solution as a power series in another variable ρ, a carefully chosen function of λ.

What we are presenting is a simple method of analytic continuation.

9.11. A Simple Algebraic Example

Let us begin with a simple idea that illustrates the foregoing discussion. Consider the problem of finding a power series development for the positive root of

$$x^2 + x = \lambda, \qquad (9.11.1)$$

where $\lambda > 0$. The explicit formula

$$x = \frac{-1 + (1 + 4\lambda)^{1/2}}{2} \qquad (9.11.2)$$

shows that the power series

$$x = \lambda - \lambda^2 + \cdots \qquad (9.11.3)$$

found by iteration, or by use of the Lagrange expansion formula, converges only for $0 \leq \lambda < \frac{1}{4}$, restraining our attention only to the λ-values of interest to us.

Perform, however, the change of variable

$$\rho = \frac{\lambda}{1 + 4\lambda}, \tag{9.11.4}$$

so that $(1 + 4\lambda) = (1 - 4\rho)^{-1}$. Then

$$x = -\tfrac{1}{2} + \frac{1}{2(1 - 4\rho)^{1/2}} = \rho + 3\rho^2 + \cdots. \tag{9.11.5}$$

The radius of convergence in the ρ-plane is also $\tfrac{1}{4}$. Hence, as long as

$$|\rho| = \left|\frac{\lambda}{1 + 4\lambda}\right| < \tfrac{1}{4}, \tag{9.11.6}$$

the series in (9.11.5) converges. This condition holds for $\lambda > 0$. Thus, (9.11.5) provides an expansion of the desired type.

EXERCISES

1. Use the identity $x^2 = \max_y (2xy - y^2)$ to obtain the representation

$$x = \min_{y \geq 0} \left(\frac{y^2 + \lambda}{1 + 2y}\right)$$

   for all $\lambda \geq 0$ for the positive solution of (9.11.1).
2. Obtain a corresponding representation for the positive solution of $x^k + x = \lambda$, for $k > 1$.
3. Obtain a corresponding representation in terms of a maximum operation for the positive solution of $x^k + x = \lambda$, $0 < k < 1$. Hence, obtain upper and lower bounds for the solution of $x^k + x = \lambda$ for $0 < k < \infty$. (*Hint:* Consider the change of variable $x^k = y$.)

9.12. The Equation $x + \lambda Bx = c$

Let us next apply the preceding ideas to the vector equation

$$x + \lambda Bx = c. \tag{9.12.1}$$

The Liouville-Neumann solution,

$$x = c - \lambda Bc + \cdots, \tag{9.12.2}$$

converges within the circle

$$|\lambda| < \frac{1}{|\lambda_M|}, \tag{9.12.3}$$

where λ_M is a characteristic root of B of largest absolute value. Without loss of generality, let us take $|\lambda_M| = 1$.

To take advantage of the fact that we are interested only in positive values of λ, let us make the change of variable

$$\rho = \frac{\lambda}{1 + \lambda}, \qquad \lambda = \frac{\rho}{1 - \rho}. \tag{9.12.4}$$

Then (9.12.1) becomes

$$x(1 - \rho) + \rho B = c - c\rho \tag{9.12.5}$$

with the power series solution

$$x = c - Bc\rho + \sum_{n=2}^{\infty} \rho^n (B - I)^n (-Bc). \tag{9.12.6}$$

The characteristic roots of $B - I$ are $\lambda_k - 1$, $k = 1, 2, \ldots$, where $\lambda_1, \lambda_2, \ldots$, are the characteristic roots of B. Hence, if B is positive definite, we see that the characteristic roots of $B - I$ are all between 0 and 1. Thus, the series in (9.12.6) converges for $|\rho| < 1$. Since this condition is satisfied for all positive λ, we have the desired enlargement of the radius of convergence.

If B is not positive definite, we may or may not obtain the desired extension, depending upon the location of the other characteristic roots of B with the absolute value 1.

EXERCISE

1. What is the situation when there is only one root, $\lambda_1 = 1$, with absolute value one with all the other roots of smaller absolute value?

9.13. The Integral Equation $f(t) + \lambda \int_0^T K(t, t_1)f(t_1)\,dt_1 = g(t)$

The reader familiar with the rudiments of the theory of Fredholm integral equations of the form

$$f(t) + \lambda \int_0^T K(t, t_1)f(t_1)\,dt_1 = g(t), \qquad (9.13.1)$$

where K is a symmetric kernel, will see readily that the foregoing techniques can be employed in the case where K is positive definite.

An advantage of this procedure, apart from possible computational purposes, is that it furnishes an explicit analytic representation of the solution of (9.13.1) for $\lambda \geq 0$ that can occasionally be useful for study of the structure of the solution.

9.14. Lagrange Multipliers

Let us now return to the study of the minimization of

$$J_a(f) = [g + Af, g + Af] + a[f, f], \qquad (9.14.1)$$

and study the dependence of the solution on the parameter a as a increases from 0 to ∞. This parameter is the Lagrange parameter we have employed previously, a "price" representing the ratio of the cost of deviation from the desired state to the cost of control.

We wish to show that

$$u_a = [g + Af_a, g + Af_a] \qquad (9.14.2)$$

is nondecreasing as a increases and that

$$v_a = [f_a, f_a] \qquad (9.14.3)$$

is nonincreasing.

What is remarkable is that these results, which are intuitively clear, can be demonstrated by a very elementary calculation. We have

$$\begin{aligned} u_a + av_a &\leq u_b + av_b, \\ u_b + bv_b &\leq u_a + bv_a, \end{aligned} \qquad (9.14.4)$$

using the fact that f_a furnishes the minimum of $J_a(f)$ and f_b the minimum of $J_b(f)$. Hence,

$$u_b + bv_b \leq u_a + av_a + (b - a)v_a \leq u_b + av_b + (b - a)v_a, \quad (9.14.5)$$

whence

$$v_b \leq v_a. \quad (9.14.6)$$

Adding the consequence of (9.14.6), $-av_a \leq -av_b$, to

$$u_a + av_a \leq u_b + av_b, \quad (9.14.7)$$

we obtain the second desired relation,

$$u_a \leq u_b. \quad (9.14.8)$$

As we shall see, the inequalities are strict provided that A^*g is not identically zero. If it is, then $f_a = 0$ for $a > 0$.

9.15. The Operator R_a

Let us now turn our attention to the properties of the operator R_a introduced in (9.8.11). There are at least three ways of deducing the structural properties of R_a. One uses standard results in Hilbert space; the second uses properties of the solution of linear integral equations; the third uses the properties of Green's functions of linear differential equations.

Since we wish to keep the discussion in this volume on a reasonably elementary level, we will merely quote the results we need as the occasion arises. What follows should be considered as an expository account of what can be obtained with the proper background in Hilbert space theory. One way of predicting the relations given below is to derive the corresponding relations for the case where A is a matrix, which is to say a finite-dimensional operator.

We begin with the observation that R_a is self-adjoint and an analytic function of a for Re $(a) > 0$. Furthermore,

$$\frac{dR_a}{da} = R_a^2, \quad \frac{d}{da} R_a^2 = 2R_a^3, \quad \frac{d}{da} f_a = R_a f_a. \quad (9.15.1)$$

One way to obtain these results is to begin with (9.11.8),

$$A^*Af_a + af_a = -A^*g \qquad (9.15.2)$$

and to differentiate with respect to a. The result is

$$(A^*A + aI)\frac{df_a}{da} = -f_a. \qquad (9.15.3)$$

This is equivalent to the third result of (9.15.1). From this, the first result follows, and the second follows similarly.

Since R_a is the inverse of a strictly negative definite operator we see that R_a is negative definite. From the relation

$$\frac{d}{da}[f_a, f_a] = 2[R_a f_a, f_a] \le 0, \qquad (9.15.4)$$

we see that $[f_a, f_a]$ is monotone decreasing as a increases. If A^*g is not identically zero, so that f_a is not identically zero, then $[f_a, f_a]$ is strictly decreasing.

From the relation

$$[g + Af_a, g + Af_a] = [g + Af_a, g] + [g + Af_a, Af_a]$$

$$= [g + Af_a, g] + [A^*g + A^*Af_a, f_a]$$

$$= [g, g] + [f_a, A^*g] - a[f_a, f_a], \qquad (9.15.5)$$

we see that

$$\frac{d}{da}[g + Af, g + Af] = [R_a f_a, A^*g] - [f_a, f_a] - 2a[R_a f_a, f_a]$$

$$= -2a[R_a f_a, f_a] \ge 0. \qquad (9.15.6)$$

Hence, $[g + Af_a, g + Af_a]$ is nondecreasing as a function of a. It is strictly increasing if f_a is not identically zero.

These results were established by means of an elementary argument in Section 9.14. However, our more detailed information concerning R_a enables us to conclude that $[f_a, f_a]$ and $[g + Af_a, g + Af_a]$ are continuous functions of a for $a > 0$.

9.16. Control Subject to Constraints

The importance of the foregoing considerations resides in the ease with which we can apply them to solve the following associated control problems:

(a) Determine $\varphi(c) = \min_f [g + Af, g + Af]$ subject to the
     constraint $[f, f] \leq c$.
(b) Determine $\psi(c) = \min_f [f, f]$, subject to $[g + Af, g +$
     $Af] \leq c$.

$$(9.16.1)$$

The first problem is that of obtaining maximum control (in this case, minimum deviation), with a limited quantity of resources. The second is that of determining the minimum cost required to maintain the cost of deviation below a certain level.

Consider first the case where $A^*g = 0$ in which case $f_a = 0$. Then

$$[g + Af, g + Af] = [g, g] + 2[g, Af] + [Af, Af]$$
$$= [g, g] + 2[A^*g, f] + [Af, Af]$$
$$= [g, g] + [Af, Af] \geq [g, g]. \qquad (9.16.2)$$

The lower limit $[g, g]$ is assumed for $f = 0$. Hence, $\varphi(c) = [g, g]$ for all $c \geq 0$. In addition, $\psi(c) = 0$ for $c \geq [g, g]$ with $\psi(c)$ undefined for $c < [g, g]$.

Consider next the nontrivial case where $A^*g \neq 0$. Since $[f_a, f_a]$ is continuous and strictly decreasing, we can introduce the quantity

$$b = \lim_{a \to 0} \{f_a, f_a\}. \qquad (9.16.3)$$

If $0 < c < b$, there is a unique $a > 0$ for which $[f_a, f_a] = c$. Then $\varphi(c)$ is determined by f_a and this is the unique maximizing function.

The proof of this last statement is the following. If there exists an $f \neq f_a$ such that $[g + Af, g + Af] \leq [g + Af_a, g + Af_a]$ with $[f, f] \leq [f_a, f_a]$, we would have

$$J_a(f) = [g + Af, g + Af] + a[f, f]$$
$$\leq [g + Af_a, g + Af_a] + a[f_a, f_a] = J_a(f_a), \qquad (9.16.4)$$

contradicting the uniqueness of the function f_a yielding the minimum value of $J_a(f)$.

To complete the discussion of $\varphi(c)$, we must consider the case where $\sup_{a\to 0} [f_a, f_a]$ is finite and c is greater than this value. Let us show in this case that there is an element f_0 to which f_a converges strongly as $a \to 0$ with the property that f_0 minimizes $[g + Af, g + Af]$ for $f \in L^2(0, T)$. Thus $\varphi(c) = [g + Af, g + Af]$ for $c \geq \sup_{a\to 0} [f_a, f_a]$.

To establish this result, let f_0 be a weak cluster point of the set of functions $\{f_a\}$. By this we mean that there is a subsequence $\{f_{a_k}\}$ converging weakly to f_0 as $k \to \infty$. Hence, $[f_0, f_0] \leq \lim_{a\to 0} [f_a, f_a]$. Since $g + Af_0$ is a cluster point of $g + Af_a$, we have

$$[g + Af_0, g + Af_0] \leq \lim_{a\to 0} [g + Af_a, g + Af_a] \leq [g + Af_a, g + Af_a].$$

$$(9.16.5)$$

If $[g + Af, g + Af] < [g + Af_0, g + Af_0]$ for some f, we would have $J_a(f) < J_a(f_a)$ for sufficiently small a, contradicting the minimal property of f_a. Hence $[g + Af_0, g + Af_0] \leq [g + Af, g + Af]$ for all f.

Furthermore, we note that $[g + Af, g + Af] = [g + Af_0, g + Af_0]$ if and only if $Af = Af_0$, as the uniqueness argument of Section 9.7 shows. If $Af = Af_0$ with $[g + Af, g + Af] \leq [g + Af_a, g + Af_a]$ for all $a > 0$, we must have $[f, f] > [f_a, f_a]$ so as not to contradict the minimality of $J_a(f_a)$. Hence,

$$[f, f] \geq \lim_{a\to 0} [f_a, f_a] \geq [f_0, f_0] \geq \lim_{a\to 0} [f_a, f_a]. \qquad (9.16.6)$$

Thus, f_0 is the unique element of least norm, as measured by $[f, f]$, in the closed linear set of functions f satisfying $Af = Af_0$.

Since this holds true for any weak cluster point, there is only one such point, namely f_0. Since $[f_a, f_a] \to [f_0, f_0]$ as $a \to 0$ and f_a converges weakly to f_0, it follows that $[f_a - f_0, f_a - f_0] \to 0$, which is to say f_a converges strongly to f_0 as $a \to 0$.

The same type of reasoning applies to the determination of $\psi(c)$ when $A^*g \neq 0$. For c in the range of $[g + Af_a, g + Af_a]$, $0 < a < \infty$, there is a unique value of a for which $[g + Af_a, g + Af_a] = c$. If c is not in the range, there are two cases: if $c < [g + Af_a, g + Af_a]$ for $a > 0$ and if $[f_a, f_a]$ is unbounded, the variational problem is meaningless, since $[g + Af, g + Af] \leq c \leq [g + Af_a, g + Af_a]$ implies $J_a(f) < J_a(f_a)$ for $[f_a, f_a] > [f, f]$. If, on the other hand, $\sup_{a\to 0} [f_a, f_a] < \infty$, then $\psi(c)$ is defined only for $c = [g + Af_0, g + Af_0]$, and $\psi(c) = [f_0, f_0]$. If $c > [g + Af_a, g + Af_a]$ for all $a > 0$, then, since $[f_a, f_a] \to 0$ as $a \to \infty$, we have $c \geq [g, g]$ and $\psi(c) = 0$, the value corresponding to $f = 0$.

9.17. Properties of $\varphi(c)$ and $\psi(c)$

Let c be in the range of $[f_a, f_a]$. Then we wish to demonstrate that

$$\frac{d}{dc}\varphi(c) = -a, \qquad (9.17.1)$$

where c and a are related by $[f_a, f_a] = c$. To establish this result, we observe that

$$\frac{d\varphi}{dc} = \left(\frac{d}{da}[g + Af_a, g + Af_a]\right)\frac{da}{dc}$$

$$= -2a[R_a f_a, f_a]\frac{da}{dc}. \qquad (9.17.2)$$

Since $c = [f_a, f_a]$, we have

$$1 = 2[R_a f_a, f_a]\frac{da}{dc}. \qquad (9.17.3)$$

Combining (9.17.2) and (9.17.3), we obtain (9.17.1).

Similarly, using the relations

$$c = [g + Af_a, g + Af_a],$$
$$\psi(c) = [f_a, f_a], \qquad (9.17.4)$$

we obtain the relation

$$\frac{d\psi(c)}{dc} = -\frac{1}{a}, \qquad (9.17.5)$$

within the range of $[g + Af_a, g + Af_a]$.

9.18. Statement of Result

Let us collect the foregoing results.

Theorem. *If $A^*g \neq 0$, either $[f_a, f_a] = c$ for some $a > 0$ with f_a the unique function providing the minimum value of $\varphi(c)$, or $[f_a, f_a] < c$ for all $a > 0$ and $f_0 = \lim_{a \to 0} f_a$ furnishes the minimum value of $\varphi(c)$.*

For c in the range of $[f_a, f_a]$, we have

$$\frac{d\varphi(c)}{dc} = -a, \qquad (9.18.1)$$

where $c = [f_a, f_a]$.

    Similarly, either $[g + Af_a, g + Af_a] = c$ for some $a > 0$ and f_a is the unique function providing the minimum value $\psi(c)$; or $[f_a, f_a]$ is bounded, $[g + Af_0, g + Af_0] = c$, and f_0 furnishes the minimum value of $\psi(c)$; or $[g + Af_a, g + Af_a] < c$ for all $a > 0$ and $f = 0$ provides the minimum value. In the range of $[g + Af_a, g + Af_a]$, we have

$$\frac{d\psi(c)}{dc} = -\frac{1}{a}, \qquad c = [g + Af_a, g + Af_a]. \qquad (9.18.2)$$

EXERCISES

1. If $g = -Ah$, then $J_a(f_a) = a[f_a, h]$.
2. Consider the problem of determining the quantity

$$\max_{f_2} \min_{f_1} \{[g + Af_1 + Af_2, g + Af_1 + Af_2]$$
$$+ a_1[f_1, f_1] - a_2[f_2, f_2]\},$$

    where $a_1, a_2 > 0$.
3. On the basis of an explicit evaluation, examine the validity of the relation

$$\max_{f_2} \min_{f_1} \{\cdots\} = \min_{f_1} \max_{f_2} \{\cdots\}.$$

Miscellaneous Exercises

1. Consider the Fredholm integral equation

$$u(x) + v(x) + \int_a^T k(x, y)u(y)\, dy = 0,$$

where $k(x, y)$ is a continuous symmetric kernel for $0 \le x, y \le T$. Show that this may be considered the Euler equation associated with the quadratic functional

$$J(u) = \int_a^T \int_a^T k(x, y)u(x)u(y)\, dx\, dy + \int_a^T u^2\, dx + 2 \int_a^T u(x)v(x)\, dx.$$

2. Show that if the quadratic part of $J(u)$ is positive definite, then the Fredholm integral equation possesses a unique solution. Write this solution in the form

$$u(x) = -v(x) + \int_a^T Q(x, y, a)v(y)\, dy,$$

where Q is symmetric in x and y.

3. Show that $Q(x, y, a)$ satisfies the Riccati-type equation

$$\frac{\partial Q}{\partial a}(x, y, a) = Q(a, x, a)Q(a, y, a).$$

4. Derive the foregoing result using the functional equation technique of dynamic programming, beginning with the representation

$$\min_u J(u) = -\int_a^T v^2\, dx + \int_a^T \int_a^T Q(x, y, a)v(x)v(y)\, dx\, dy$$

$$= f(v(x), a),$$

where f is a quadratic functional of v.

(See

R. Bellman, "Functional Equations in the Theory of Dynamic Programming—VII: A Partial Differential Equation for the Fredholm Resolvent," *Proc. Amer. Math. Soc.*, **8**, 1957, pp. 435–440.)

5. Consider the second-order linear differential equation

$$L(u) = u'' + a(t)u = f(t), \qquad u(0) = u(T) = 0,$$

with the solution

$$u = \int_0^T k(t, s)f(s)\, ds,$$

where k is the Green's function. Write $u = T(f)$. Let

$$(u, v) = \int_0^T uv\, dt,$$

and let T^*, the adjoint operator, be defined by $(T(u), v) = (u, T^*v)$, so that

$$T^*(v) = \int_0^T k(t, s)v(t)\, dt.$$

Use the relations

$$\left(L\left(\int_0^T k(t,s)f(s)\,ds\right),g\right)=(f,g)=\left(\int_0^T k(t,s)f(s)\,ds,L(g)\right)$$

$$=(f,T^*(L(g)))=\left(f,\int_0^T k(t_1,t)L(g)\,dt_1\right),$$

to show that $k(t_1,t)=k(t,t_1)$.

6. If

$$u(x,y)\le c+\int_0^x\int_0^y v(r,s)u(r,s)\,dr\,ds,$$

where $c\ge 0$, $u(r,s)$, $v(r,s)\ge 0$, then

$$u\le c\exp\left(\int_0^x\int_0^y v(r,s)\,dr\,ds\right).$$

7. If

$$u(x,y)\le a(x)+b(y)+\int_0^y\int_0^x v(r,s)u(r,s)\,dr\,ds,$$

then

$$u\le\frac{[a(0)+b(y)][a(x)+b(0)]\exp\left(\int_0^y\int_0^x v\,dr\,ds\right)}{a(0)+b(0)}$$

under appropriate positivity assumptions.

8. If

$$u(x,y)\le c+a\int_0^x u(x,y)\,ds+b\int_0^y u(x,s)\,ds,$$

then

$$u(x,y)\le c\exp[ax+by+abxy],$$

under appropriate positivity assumptions.

9. If

$$u(x,y)\le a(x)+b(y)+a\int_0^x u(x,y)\,ds+b\int_0^y u(x,s)\,ds,$$

then $u(x,y)\le Q(x,y)$, where

$$Q(x,y)=\frac{\begin{aligned}&[a(0)+b(0)+\int_0^y e^{-by_1}b'(y_1)\,dy_1][a(0)+b(0)\\&+\int_0^x e^{-ax}a'(x_1)\,dx_1]\exp[ax+by+abxy]\end{aligned}}{[a(0)+b(0)]}.$$

BIBLIOGRAPHY AND COMMENTS

9.1. The results of this chapter first appeared in

R. Bellman, I. Glicksberg, and O. Gross, "On Some Variational Problems Occurring in the Theory of Dynamic Programming," *Rend. Circ. Mat. Palermo*, **3**, 1954, pp. 1–35.

The results concerning Hilbert space and resolvent operators may be found in

F. Riesz and B. Sz. Nagy, *Functional Analysis* (English translation), Ungar, New York, 1955,

and in

R. Bellman, I. Glicksberg, and O. Gross, *Some Aspects of the Mathematical Theory of Control Processes*, The RAND Corporation, **R-313**, 1958.

9.11. This discussion follows

R. Bellman, "On Perturbation Methods Involving Expansions in Terms of a Parameter," *Quart. Appl. Math.*, **13**, 1955, pp. 195–200.

AUTHOR INDEX

241

SUBJECT INDEX